DISASTER RESILIENT CITIES

T0282380

DISASTER RESILIENT CITIES

Concepts and Practical Examples

YOSHITSUGU HAYASHI
YASUHIRO SUZUKI
SHINJI SATO
KENICHI TSUKAHARA

AMSTERDAM • BOSTON • HEIDELBERG • LONDON
NEW YORK • OXFORD • PARIS • SAN DIEGO
SAN FRANCISCO • SINGAPORE • SYDNEY • TOKYO

Butterworth-Heinemann is an imprint of Elsevier

CONTRIBUTORS

Y. Akiyama
Center for Spatial Information Science, The University of Tokyo, Chiba, Japan

Y. Hayashi
Graduate School of Environmental Studies, Nagoya University, Nagoya, Japan

T. Inamura
The Open University of Japan, Chiba, Japan

S. Ishii
Graduate School of Environmental Studies, Nagoya University, Nagoya, Japan

N. Kachi
Graduate School of Engineering, Kyushu University, Fukuoka, Japan

H. Kato
Graduate School of Environmental Studies, Nagoya University, Nagoya, Japan

T. Kato
Institute of Industrial Science, The University of Tokyo, Tokyo, Japan

Y. Kawazoe
Institute of Industrial Science, The University of Tokyo, Tokyo, Japan

N. Matsuta
Graduate School of Education, Okayama University, Okayama, Japan

A. Mimuro
(former) Graduate School of Environmental Studies, Nagoya University, Nagoya, Japan

S. Nakamura
Department of Civil Engineering, Nagoya University, Nagoya, Japan

Y. Ogawa
Graduate School of Frontier Sciences, The University of Tokyo, Chiba, Japan

S. Sato
Department of Civil Engineering, The University of Tokyo, Tokyo, Japan

R. Shibasaki
Center for Spatial Information Science, The University of Tokyo, Chiba, Japan

T. Shimozono
Department of Civil Engineering, The University of Tokyo, Tokyo, Japan

K. Sugimoto
Graduate School of Environmental Studies, Nagoya University, Nagoya, Japan

N. Sugito
Faculty of Humanity and Environment, Hosei University, Tokyo, Japan

Y. Suzuki
Disaster Mitigation Research Center, Nagoya University, Nagoya, Japan

Y. Tajima
Department of Civil Engineering, The University of Tokyo, Tokyo, Japan

H. Tanikawa
Graduate School of Environmental Studies, Nagoya University, Nagoya, Japan

K. Tsukahara
Graduate School of Engineering, Kyushu University, Fukuoka, Japan

M. Yoshitake
Institute of Industrial Science, The University of Tokyo, Tokyo, Japan

Butterworth-Heinemann is an imprint of Elsevier
The Boulevard, Langford Lane, Kidlington, Oxford OX5 1GB, UK
50 Hampshire Street, 5th Floor, Cambridge, MA 02139, USA

Notices
Knowledge and best practice in this field are constantly changing. As new research
and experience broaden our understanding, changes in research methods, professional
practices, or medical treatment may become necessary.

Practitioners and researchers must always rely on their own experience and knowledge
in evaluating and using any information, methods, compounds, or experiments described
herein. In using such information or methods they should be mindful of their own
safety and the safety of others, including parties for whom they have a professional
responsibility.

To the fullest extent of the law, neither the Publisher nor the authors, contributors, or
editors, assume any liability for any injury and/or damage to persons or property as a
matter of products liability, negligence or otherwise, or from any use or operation of any
methods, products, instructions, or ideas contained in the material herein.

Library of Congress Cataloging-in-Publication Data
A catalog record for this book is available from the Library of Congress

British Library Cataloguing-in-Publication Data
A catalogue record for this book is available from the British Library

ISBN: 978-0-12-809862-2

For information on all Butterworth Heinemann publications
visit our website at https://www.elsevier.com/

Working together
to grow libraries in
developing countries

www.elsevier.com • www.bookaid.org

CONTENTS

PREFACE

Since the beginning of the 21st century, the world has experienced many natural disasters of a type and scale never before experienced in the affected areas. Many people have lost their lives and property, while logistical routes have been cut, causing enormous damage to industry. A massive disaster may deal a heavy blow to the country in an instant, devastating its economy and society irreversibly.

Focusing on irreversibility, the term "resilience" has emerged as an important concept for the nation and for regional communities. Resilience refers to the ability to quickly recover the balance that is temporarily lost. It is a concept that also comprises "preparation," which helps prevent damage from reaching a fatal, irreversible level. It is therefore important to note that resilience, which is the ability to recover from a short-term impact, is a prerequisite to ensuring "sustainability," the ability to maintain the economy, environment, and society of the region in a long term. Imagine a scene in which a circus tightrope walker standing on a rope loses his balance for a moment. Unless he recovers his balance immediately, he cannot take the next step forward. If he can recover a stable footing at each step, he can keep walking forward step by step. To restore his balance in a moment, he has to hold a long pole. What serves as the balancer that will protect regional communities from a fatal blow if they encounter a massive disaster? The goal of this book is to answer this question.

This question is reflected in the keyword "resilience," and the awareness of this issue should be shared in the international community. Poverty and insufficient embankments, as well as problems in transportation and other infrastructure, make developing countries far more vulnerable than developed countries.

This book examines various disasters that have occurred in various regions with diverse cultural and economic backgrounds, and explores the conditions for resilience against them. Focusing particularly on the Great East Japan Earthquake, a massive disaster that took place in Japan, home to these authors, its natural influence is mapped, and physical damage is captured and described from multiple angles. Finally, this book presents an estimate of the quality of life, as the ultimate impact on people's lives.

The Great East Japan Earthquake in 2011 revealed that Japanese society, which has undergone modernization and entered a mature stage, is extremely vulnerable to a disaster. The exclusion of a large tsunami from the assumptions for formulating disaster measures resulted in huge damage, injuries, and loss of life, and even a serious accident at a nuclear power plant. Despite the lesson of 2008, when people around the world were shocked by news footage of the tsunami in the Indian Ocean, we learned nothing, never imagining it would happen to us. Although some ancient writings describe a large tsunami that occurred several hundred years ago, we see it as just an event in the past. Although such information exists and is disseminated, people do not comprehend it properly. The lack of both awareness of events that occurred in the distant past and the feeling that they might occur again is a major fault of modern society. To acquire this ability, we need a mechanism that enables us to feel and share the threats of nature by constantly integrating both traditional knowledge and the latest information from around the world. This book presents a method developed by applying big data for this purpose.

It is important to note that resilience varies among countries or regions depending on their natural and cultural backgrounds, before examining the factors that may lower resilience and exploring measures for its efficient recovery or improvement. To this end, we must learn from traditional knowledge first, listen simply to the voice of nature, and understand local climates and cultures, thereby humbly seek the most appropriate way without overconfidence in science and technology. The ultimate goal of enhancing resilience is to maintain and improve the quality of life, or happiness, for people, and therefore its effects should be measured from this viewpoint.

We organize the lessons of the Great East Japan Earthquake and also learn from overseas cases. In this book, we discuss on this issue from various viewpoints, including civil engineering, urban engineering, geography, and cultural anthropology, and present a concept of resilience revealed through such discussions, which can be shared (or complemented) across disciplinary fields.

Specific measures to enhance resilience should be incorporated in national or regional land use planning, which include organizing hazard information, building hazard-conscious communities, and securing engineers and heavy machinery for recovery/reconstruction work. Especially in future land use or infrastructure development planning, it is necessary to measure resilience and sustainability, and to incorporate them in prior assessment for the plans. To this end, a set of environmental information called Geo Big Data is useful. By analyzing this data, disaster prevention/mitigation effects can be assessed in advance, enabling the government to select appropriate national land design policies. After a policy is put into force, monitoring its effects is necessary. This book presents the methodology for this based on various study cases.

This book is a summary of the results of "GRENE-City: designing cities and national land as resilient—an application of environment information technology" (representative: Yoshitsugu Hayashi, 2011–15), one of the research topics in the environmental information category of the Green Network of Excellence (GRENE) program of the Ministry of Education, Culture, Sports, Science and Technology.

The editors are indebted to Professor Toshio Koike for his excellent leadership in the whole research project, Dr. Kenji Sugimoto and Ms. Naoko Dainobu for their substantial assistance.

Yoshitsugu Hayashi
Yasuhiro Suzuki
Shinji Sato
Kenichi Tsukahara

CHAPTER 1

Introduction

KEY MESSAGE

Since the beginning of the 21st century, as changes in the global environment and social structures have increased the threat of natural disasters, the significance of resilience has been attracting global attention. Tracing the global-scale changes from the 1990s reveals why resilience is needed now and what hope the term "resilience" represents. To improve resilience, it is important to clarify the causes and mechanisms that may lower resilience, and to design specific measures to halt their progression.

CHAPTER 1.1

Introduction—Why Is Resilience Lost?

Y. Suzuki*, Y. Hayashi†, K. Tsukahara‡
*Disaster Mitigation Research Center, Nagoya University, Nagoya, Japan
†Graduate School of Environmental Studies, Nagoya University, Nagoya, Japan
‡Graduate School of Engineering, Kyushu University, Fukuoka, Japan

1.1.1 RESILIENCE NOW GAINING ATTENTION

Resilience is originally a concept that began to be used in psychology and social ecology in the 1970s, meaning the ability to recover from a negative event. As many natural disasters have occurred since the beginning of the 21st century, such as the Indian Ocean earthquake off the coast of Sumatra, Indonesia, large-scale hurricanes in the United States, super-typhoons in the Philippines, the Great East Japan Earthquake and large tsunami in Japan, and the massive flooding in Bangkok, the term has been used in a broader sense in association with disasters. At many international organizations and government agencies, "disaster resilience" is becoming a keyword [1–4].

The Millennium Development Goals, or MDGs, eight objectives for the international community in the 21st century established by the United Nations in 2000, did not include disaster prevention. In response to the subsequent international efforts to place a greater focus on disaster prevention and a series of massive disasters that took place in the 2000s, however, some objectives related to disaster prevention or disaster mitigation were set in the proposal for the Sustainable Development Goals, or SDGs, submitted to the president of the UN General Assembly in Aug. 2015 [5]. Among them, "Goal 11. Make cities and human settlements inclusive, safe, resilient and sustainable" is particularly noteworthy. The SDGs, setting the improved resilience in cities and the human living environment as an objective, were finally adopted at the UN Summit in Sep. 2015.

The Japanese government emphasizes the importance of national resilience. One reason for this is that though Japan had implemented various disaster countermeasures under the Disaster Countermeasures Basic Act and several other relevant laws since the Ise Bay Typhoon in 1959, the inadequacy of these measures in responding to the Great East Japan Earthquake, a large-scale disaster, demonstrated Japan's lack of ability to achieve recovery

Disaster Resilient Cities
http://dx.doi.org/10.1016/B978-0-12-809862-2.00001-2

and reconstruction smoothly. The fundamental plan for national resilience, titled "Building national resilience—creating a strong and flexible country," [6] was approved by the cabinet on Jun. 3, 2014 and emphasizes two points in its basic concept: (1) assuming the possibility of the worst cases without preconceptions; and (2) preparing comprehensive measures incorporating national land policy and industry policy, not limited to the scope of narrowly defined disaster prevention. As reported by the Fukushima Nuclear Accident Independent Investigation Commission of the National Diet of Japan, the Great East Japan Earthquake was not a disaster of a kind that had never occurred before. Similar tsunamis that followed the Jougan Earthquake in 869 and the Keicho-Sanriku Earthquake in 1611 were excluded from the assumptions for formulating disaster measures, resulting in huge damage, even causing an accident at a nuclear power plant [7]. Reflecting regret for this, the fundamental plan above declares preventing the same mistake from being made again as the primary goal of national resilience.

It is necessary to observe calmly the situation of society today and in the near future, and review the whole picture of national land structures and social preparedness, which could not be captured within the conventional framework of disaster prevention. In this sense, resilience has been increasingly seen as important.

1.1.2 WHY RESILIENCE HAS BECOME AN ISSUE NOW

In Japan, the concept of resilience first attracted public attention in the aftermath of the Kobe Earthquake in 1995. This earthquake provided a major turning point requiring a review of Japan's disaster prevention systems at the end of the 20th century.

The Kobe Earthquake was caused by movement of an active fault that lies from Awaji Island to around Kobe. The presence of this active fault became known in the 1970s and it was even designated as a fault that requires precautions in the early 1980s [8,9], though it was not included as a target of disaster prevention measures. Before 1995, because of the lack of active provision of information on hazards, most residents were unaware of the risk of the active fault and believed that a major earthquake would not take place in their living area. People were, therefore, shocked when they saw modern buildings and expressways collapse. The "myth of safety" was destroyed. It was a painful lesson from which people learned that although the postwar efforts had prepared Japan sufficiently for small- or medium-scale, highly frequent natural disasters, such as typhoons, the country's preparedness for low-frequency, high-consequence disasters was totally insufficient. Recognizing the inability to prevent disasters completely resulted in the emergence of an approach that placed greater emphasis on "disaster mitigation" than on "disaster prevention." After this earthquake, people began to doubt the earthquake resistance of nuclear power plants.

The Great Sumatra Earthquake in 2004 took place in the sea around the Andaman and Nicobar Islands, which was an area where a major earthquake was unlikely to occur according to the general plate tectonics theory. It was a huge earthquake with a magnitude of between 9.1 and 9.3. A large tsunami hit the coast of the Indian Ocean, leaving deaths and the missing totaling over 170,000 in Indonesia, 40,000 in Sri Lanka, 20,000 in India, and 5,000 in Thailand. Images of the tsunami devastating tourism spots and cities shocked people around the world. Entering the 21st century, we have seen global warming and other environmental challenges escalating. Signs of severer weather disasters, such as sea level rise and torrential rainstorms, have been increasingly observed worldwide. Hurricane Katrina in 2005 killed nearly 2,000 people in the United States, while tornados, snowstorms, and droughts took place all around the world. Rainstorms and floods began to occur frequently in Europe.

Under these circumstances, in Japan, a sense of fear of large-scale natural disasters also began to rise around 2005. The authors, Hayashi and Suzuki, as part of our activities on the Science Council of Japan, gathered academic information concerning natural disasters and in 2007 formulated a proposal regarding the basic idea of mitigating damage from natural disasters in the future (Policies for the creation of a safe and secure society in light of increasing natural disasters around the World) [10]. This proposal presented as the policies that the Japanese government should take: (1) a paradigm shift from a "short-term perspective focusing on economic growth" to "long-term creation of safe and secure societies"; (2) infrastructure improvement based on a consensus on the level of disaster safeguard required; (3) reform of national land use from a long-term perspective; (4) well-balanced use of physical and management countermeasures; (5) reform of depopulated areas based on vulnerability assessment and recognition of potential risks; (6) establishment of integrated disaster management policies and strategies by the national and municipal governments; (7) development of public awareness of and preparedness for disasters by utilizing hazard maps; (8) promotion of education for natural disaster mitigation; (9) promotion of activities by NPOs and NGOs; (10) promotion of international cooperation in disaster mitigation; and (11) building sustainable strategies and taskforces for natural disaster mitigation.

The proposal above officially pointed out for the first time in Japan that vulnerability to disasters is an issue that influences the entire structure of society and national land, and that vulnerability has been increasing in modern society, which is often considered safer than before. It is particularly noteworthy that the proposal presented as the first policy "paradigm shift" from "short-term perspective focusing on economic growth" to "long-term creation of safe and secure societies."

Unfortunately, the 2011 Great East Japan Earthquake took place before this proposal was effectively employed in establishing and implementing national or municipal policies.

1.1.3 ISSUES ON RESILIENCE RAISED BY THE GREAT EAST JAPAN EARTHQUAKE

Among many issues raised by the Great East Japan Earthquake, the four issues below are particularly significant for resilience in terms of social preparedness.

First is the issue of "exclusion from assumptions." While seismologists were unable to predict a massive earthquake with a magnitude of 9.0 in the Japan Trench, geology and history recognized that a tsunami had swept deep into the Sendai Plain, causing serious damage, in the aftermath of the Jougan Earthquake in 869 and the Keicho-Sanriku Earthquake in 1611. The Headquarters for Earthquake Research Promotion therefore announced in 2004 that a large interplate earthquake could occur along the Japan Trench off Miyagi and Fukushima prefectures. The Central Disaster Management Council of the Cabinet Office, however, decided not to consider this possibility in disaster countermeasures, and Tokyo Electric Power Company and the Nuclear and Industrial Safety Agency followed this decision.

The regulatory guide for reviewing seismic design of nuclear power reactor facilities established by the Nuclear Safety Commission of Japan in 2006 includes a provision: "Safety functions of the Facilities shall not be significantly impaired by tsunami which could be reasonably postulated to hit in a very low probability in the service period of the Facilities." This sentence allowed an interpretation that measures are not necessarily to be taken if a tsunami is scientifically likely to hit the facilities but such a tsunami is judged not to be reasonably postulated [11].

Information on the risk of a large earthquake is often an "inconvenient truth," considering its social impact. While failure to scientifically predict a M9.0 earthquake is generally seen as the major cause of the accident at the Fukushima Daiichi Nuclear Power Station, the social weakness in the ability to assume an inconvenient truth (failure to accept the presence of an inconvenient truth and decide to take proper action) was actually the biggest problem. We understand that regret for this is reflected in the statement of "assuming the possibility of the worst cases without preconceptions" included in the fundamental plan for national resilience mentioned previously. The attitude of always seeking "long-term reasonableness in view of future generations" without turning away from inconvenient truths is required for disaster countermeasures from now on.

Second is the issue of vulnerability underlying the national land structure and social situation. As resilience has the meaning of pliable flexibility, it is important not only to be prepared directly for a purpose, but also to take indirect measures so as to secure some allowance, or margin of error, in the measures. A system developed through the excessive pursuit of rationalization, placing top priority on economic efficiency, will collapse when it exceeds an anticipated level, leaving no allowance. To enhance resilience, our "short-term economic efficiency first" policy should be abandoned. This is not easy, though it is very important.

Modern society, developed through the pursuit of convenience, is vulnerable to disasters. Cell phone networks will become congested in the event of a disaster, as there is no redundancy in the system. The conventional attitude of trying to construct towns and buildings even in places that are vulnerable to disasters by fully utilizing technological capabilities should also be reviewed. If locational conditions are bad, high maintenance and management cost is required, undermining long-term sustainability. It is important to take a well-balanced approach by making use of the allowance created not only through pushing (protecting with infrastructure) but also pulling (withdrawing from use of the land).

To achieve this, as suggested by the aforementioned proposal of the Science Council of Japan in 2007, a paradigm shift is necessary. Withdrawal from the use of land may generate conflict with individual property rights or vested interests and therefore requires strong political leadership and high public awareness of resilience. Even policies that are not associated with short-term economic benefits should be carefully examined from the perspective of long-term reasonableness, taking into consideration the benefits and losses for future generations. If they are considered necessary, the results of examination should be presented to the public and the efforts should be made to obtain social consensus on well-balanced investment allocation for the future. Academia must construct theories and conduct verifications based on objective data to support this approach. The specific methods of developing policies to achieve targets based on long-term reasonableness are presented in Chapter 4.

The third issue is a lack of comprehensive understanding of linked or complex phenomena. In the case of the Great East Japan Earthquake, subsidence occurred in the coastal land before the area was hit by tsunami, resulting in worsening of damage. If the ground-supporting capacity of sea embankments is lowered due to soil liquefaction, they will collapse easily when hit by a tsunami. As a result of a tsunami running up rivers, towns were hit by the tsunami from behind. The tsunami also destroyed the petroleum tanks on the coast, causing fires. The chain of disasters stemming from the nuclear plant accident associated with radiation leakage was beyond our imagination. Moreover, problems in society, such as inappropriate land use, aging, and weak community capabilities, may aggravate the situation after the disaster. The mechanisms of these complex phenomena have not been sufficiently considered in the studies and assumptions for disaster countermeasures in the past.

Estimation of damage conducted by the government as the first step in developing disaster countermeasures tends to focus only on the total amount of damage. While the numbers of houses and buildings destroyed or people killed according to the magnitude of an earthquake are roughly estimated based on past experience, accumulation of data on individual damage or detailed simulation of the mechanism of damage occurrence are not included in the estimation. Without these activities, the bottleneck that lowers resilience cannot be identified, and the direction of measures cannot be clarified. We must start by archiving and analyzing details of the Great East Japan Earthquake. So-called "big data," including various kinds of digital national land information and human behavioral records, will be significantly useful in disaster simulation. It is important to conduct simulations fully incorporating diverse natural disasters, thereby identifying factors that may cause loss of resilience and the measures to recover it.

Fourth is the issue of autonomy of residents in disaster prevention/mitigation. Emphasis on self-help and mutual help is not only for the purpose of compensating for the limitations of public assistance. Evacuation is the act of preparation of primary importance to protect human lives, rather than an act that is taken just because the embankment is insufficient. Whether the residents understand the importance of voluntary evacuation and mutual assistance and can take autonomous action affects the basis of resilience.

Mental factors such as motivation and a feeling of happiness are also important for resilience. It is necessary to develop a framework that pays respect to ethnic or local traditional knowledge and cultures and encourages young people especially to take voluntary action. Conventional disaster prevention measures, lacking consideration of these points, have been devoted to constructing uniform, standardized embankments, enhancing aseismic performance, and developing warning systems for residents and risk management systems for government agencies.

As so far described, the Great East Japan Earthquake has presented many challenges related to resilience to us. We must comprehensively review the entire structure of national land and society, taking into consideration that they are all deeply related to resilience. In this process, it is particularly important to face severe hazards or inconvenient truths while aiming to realize a world where many people feel happy. Specific policies to ensure a long-term sustainable quality of life are required.

REFERENCES

[1] UN/ISDR (United Nations International Strategy for Disaster Reduction Secretariat). Hyogo framework for action 2005–2015: building the resilience of nations and communities to disasters. http://www.unisdr.org/2005/wcdr/intergover/official-doc/L-docs/Hyogo-framework-for-action-english.pdf; 2007 [accessed 01.09.15].

[2] Council of Australian Governments. National strategy for disaster resilience, Attorney-General's department. https://www.ag.gov.au/EmergencyManagement/Documents/NationalStrategyforDisasterResilience.PDF; 2011 [accessed 01.09.15].

[3] FEMA (Federal Emergency Management Agency). Crisis response and disaster resilience 2030: forging strategic actions in an age of uncertainty. http://www.fema.gov/media-library-data/20130726-1816-25045-5167/sfi_report_13.jan.2012_final.docx.pdf; 2012 [accessed 01.09.15].

[4] The World Bank. Building resilience—integrating climate and disaster risk into development. http://www.worldbank.org/content/dam/Worldbank/document/SDN/Full_Report_Building_Resilience_Integrating_Climate_Disaster_Risk_Development.pdf; 2013 [accessed 01.09.15].

[5] United Nations. Sustainable development goals. United Nations. http://www.un.org/sustainabledevelopment/sustainable-development-goals; 2015 [accessed 01.09.15].

[6] National Resilience Promotion Office, Cabinet Secretariat. Building national resilience—creating a strong and flexible country. http://www.cas.go.jp/jp/seisaku/kokudo_kyoujinka/index_en.html; 2014 [accessed 01.09.15].

[7] The National Diet of Japan. The official report of the Fukushima Nuclear Accident Independent Investigation Commission. http://warp.da.ndl.go.jp/info:ndljp/pid/3856371/naiic.go.jp/en/report; 2012 [accessed 01.09.15].

[8] Matsuda T. Active faults and damaging earthquakes in Japan—macroseismic zoning and precaution faults zones. In: Simpson DW, Richards PG, editors. Earthquake prediction. Maurice Ewing Series, vol. 4. Washington, DC: American Geophysical Union; 1981. p. 279–89.

[9] Suzuki Y. Tectonic geomorphological active fault studies in Japan after 1980. Geogr Rev Jpn B 2013;86(1):6–21. https://www.jstage.jst.go.jp/article/geogrevjapanb/86/1/86_860101/_article [accessed 01.09.15].

[10] Science Council of Japan. Policies for the creation of a safe and secure society in the light of increasing natural disasters around the world, http://www.scj.go.jp/ja/info/kohyo/pdf/kohyo-20-t38-4e.pdf; 2007 [accessed 01.09.15].

[11] Suzuki Y. Nuclear power plants and active faults—never allow disasters caused by "exclusion from assumption". Iwanami Shoten, Tokyo; 2013.

CHAPTER 2

Emerging Crisis in Resilience over the World

KEY MESSAGES

- Characteristics of the 2011 Tohoku Tsunami were described on the basis of collaborative tsunami damage surveys performed by the Joint Survey Group. The complex behaviors of the mega-tsunami were characterized by the magnificent scale and the low occurrence frequency. Emerged risks in the hazard scale, the initial responses, and the restoration processes appeared to provide valuable lessons for enhancing resilience of coastal communities.
- Under international cooperation, we implemented an on-site investigation and established disaster data archives regarding the storm surge caused by super typhoon Haiyan that hit the Philippines in Nov. 2013. Based on these activities, we clarified the relationship between the geography of coral reefs and characteristics of flood damage. We also examined what must be done to improve resilience in coastal lowlands of developing countries.

- Chapter 2.3 describes the cause and effect of the 2011 Thai flood and the measures actually taken based on on-site investigation. The 2011 Thai flood inflicted serious damage not only on economics but also on the daily life due to the prolonged innundation period which is a characteristic of a delta region. Our research group estimates the effect on daily life in the Bangkok Metropolitan Region (BMR) based on the changes in accessibility for urban facilities and clarifies the daily life in the newly developed areas in the northern part of BMR, which was affected the most.

- Chapter 2.4 takes as an example the damage and recovery from an earthquake and tsunami that occurred in Peru in 2007 and discusses natural disasters as social and cultural phenomena. Specifically, this section will discuss the vulnerability of the urban area of Peruvian cities and its historical background (conquest and colonial occupation) compared with the resilience in the ancient era of Andean civilization and sustainability of indigenous communities that partially continued the resilience of their ancestors.

- Mongolian nomads have been always living with nature, and have coped effectively with natural disasters since ancient times. Traditional wisdom to live a sustainable and resilient life remains in their nomadic lifestyle. By contrast, while modern civilized urbanization with over-concentration in Ulaanbaatar has progressed rapidly in recent years, various urban problems have obviously surfaced. Especially in the *ger* districts, where many *gers* used by nomads stand, there are various environmental problems; therefore, redevelopment projects have recently been proceeding. In this section, we review the traditional significance of those *ger* districts for Mongolian people, study the consciousness of residents about the redevelopment, and discuss the prospect of the project.

- The resilience of communities in Japan has been supported by wooden houses constructed with materials and techniques available in the neighboring areas. However, in recent years, production capacity of wooden houses has been decreasing nationally. In Chapter 2.6, using field surveys and various statistics, we would like to clarify two structural issues: (1) decline of the production network in communities; and (2) decrease of materials and technicians, which have been brought about by the change in wooden-house production in the postwar period. Using specific cases, we would also like to study the need for wooden-house production capacity that surfaces after a disaster, and issues that should be dealt with in ordinary times.

CHAPTER 2.1

The Great East Japan Earthquake

S. Sato
Department of Civil Engineering, The University of Tokyo, Tokyo, Japan

2.1.1 CHARACTERISTICS OF THE 2011 TOHOKU TSUNAMI

The earthquake that occurred off the Tohoku coast in the Pacific at 2:46 pm JST on Mar. 11, 2011 and had the moment magnitude scale of $M_w = 9.0$ shook extensive areas of Japan and triggered a giant tsunami. The tsunami struck many coastal areas on the Pacific coast of eastern Japan, and reached lands at a height of a maximum 40 m, and did catastrophic damage to many areas there. As of Aug. 2015, more than 18,000 people had been found dead or were still missing. The Tohoku region, which suffered serious damage from the last tsunami, has repeatedly been struck by tsunamis, including the Meiji Sanriku Tsunami in 1896 and the Showa Sanriku Tsunami in 1933. Therefore, comprehensive measures and programs, including structural measures, such as seawalls, that prevent land from being flooded by tsunamis, and nonstructural measures that mainly consist of quick evacuation plans, had been introduced to the region; Tohoku was the most prepared in the world against tsunamis. The introduced measures and programs were effective in reducing damage done by the Tohoku Tsunami, but the world was astonished at how limited the effectiveness was when the giant tsunami, far larger than the tsunamis which structural measures were designed to block, took place, and shocked at how many people were killed in the tsunami.

It is important to understand correctly the characteristics of the tsunami in order to examine how to rebuild damaged areas and communities there, but unfortunately there are not sufficient scientific data available, except data from tide gages in some locations. For this reason, it is important to study watermarks left immediately after the tsunami. A survey of watermarks left by tsunamis needs to be carried out efficiently to cover areas damaged by the tsunami quickly, before the marks are gone. On the day after the tsunami caused by the 2011 earthquake that occurred off the Pacific coast of Tohoku (hereafter called the "Tohoku Tsunami"), a few academic organizations jointly set up a website to share information, and successfully carried out the joint survey efficiently based on a unified manner of sharing

Hakozaki-Shirahama, Kamaishi, Iwate

Tsunami watermark
T.P.+14.1m

Seawall height
T.P.+5.6m

Fig. 2.1.1 A typical example of a tsunami watermark survey in Kamaishi, Iwate Prefecture (Apr. 11, 2011).

information. Fig. 2.1.1 is an example of a tsunami survey on Kamaishi City, Iwate Prefecture. In this city, houses and land there were flooded by the tsunami, which overtopped 5.6-m-high seawalls. Judging from wreckage carried by water, the tsunami was confirmed to have reached a point at an altitude of 14.1 m. A survey of this kind was carried out nationwide to study watermarks in the areas hit by tsunamis. Access to some areas of Fukushima Prefecture, where there was an accident at the Fukushima Daiichi Nuclear Power Station, was still restricted, but by 1 year after the accident, almost all areas damaged by the tsunami, including the restricted areas, were surveyed to examine watermarks.

Fig. 2.1.2 shows a comparison between the measurements by the joint survey group of the altitude of the watermarks left by the Tohoku Tsunami [1] and those of the watermarks left by previous tsunamis [2]. It can be confirmed that areas from Hokkaido to the Boso Peninsula observed high tsunamis caused by the 2011 Tohoku Tsunami, and that the tsunami was particularly high in the Sanriku region and reached as high as 40 m above sea level in some areas of the region. Photos taken during the research by the joint survey group have been uploaded as archives [3] on Web Geographic Information

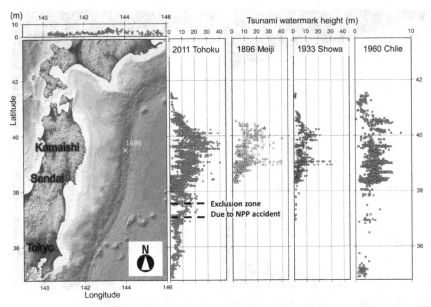

Fig. 2.1.2 Distribution of tsunami watermark heights compared with historical tsunamis. *(Data from TTJS (Tohoku Tsunami Joint Survey Group), http://www.coastal.jp/tsunami2011, and Japan Tsunami Trace Database, http://tsunami-db.irides.tohoku.ac.jp [accessed Aug. 2015]).*

System (GIS), open to anyone. A comparison with previous tsunamis reveals that the Tohoku Tsunami was as high as the Meiji Sanriku Tsunami but traveled several times farther; this tsunami is confirmed to be one of the largest in terms of the height and distance it reached. Later studies of seismic waves and source zones of the Tohoku Tsunami suggest that the tsunami was a combination of two different tsunamis that occurred at the same time in response to an extremely giant earthquake: one is the Sanriku Tsunamis, which repeat once in a 100 years or so (their recent examples are the 1896 Meiji Sanriku Tsunami and the 1933 Showa Sanriku Tsunami), and the other is an extensive tsunami which is considered to repeat once in about a 1000 years (their recent example is the 869 Jougan Earthquake and Tsunami) [4].

On the basis of extensive tsunami surveys combined with limited number of tide gage and buoy data, the tsunami source was investigated. When the tsunami source is given, the propagation of tsunami in the ocean can be simulated on the basis of the long wave equations. Fig. 2.1.3 illustrates the results of numerical computations of the 2011 Tohoku Tsunami performed by Sato et al. [5]. Fig. 2.1.3A shows a tsunami snapshot 6 min after the earthquake and Fig. 2.1.3B shows one 28 min after the earthquake. Such numerical simulation is found to be useful in the design of seawalls and evacuation plans.

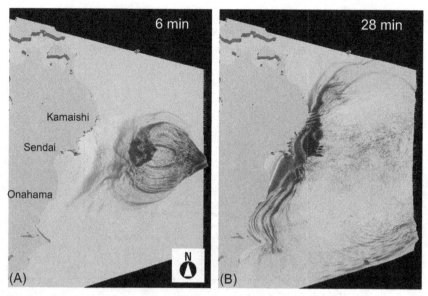

Fig. 2.1.3 (A,B) Numerical computation of the 2011 Tohoku Tsunami.

2.1.2 CHARACTERISTICS OF THE DAMAGE DUE TO TSUNAMI

2.1.2.1 High Run-Up Height and Extensively Flooded Areas

Characteristics of tsunamis are that they become higher as the water depth becomes shallower, and they tend to become higher behind the shallows like a convex lens. Tsunamis also tend to grow higher at the landside end of a V-shaped bay or a port toward which the bay or port becomes gradually narrower. The bathymetry contour lines of the Sanriku coast project convex toward the sea and the Sanriku coastline forms rias with many V-shaped bays, and past tsunamis demonstrated that areas with such bays or in such landforms were subject to severe damage. As Fig. 2.1.2 shows, the 2011 Tohoku Tsunami was confirmed to have left watermarks in a high place of the complicated sawtooth coast stretching from central to northern Iwate Prefecture. Another characteristic of the 2011 Tohoku Tsunami is that it also flooded a large area of lowlands near the mouth of small rivers and large plains in Sendai and Ishinomaki cities. Many such lowlands, including those of Ishinomaki City, Miyagi Prefecture and Rikuzentakata City and Otsuchi Town, Iwate Prefecture, experienced a rapid population increase when the country enjoyed remarkable economic growth after the Second World War. Although seawalls had been built there according to the record of past tsunamis, the Tohoku Tsunami overtopped the walls and claimed the lives of many people there.

2.1.2.2 Delay in Evacuation and Unexpectedly Extensively Flooded Areas

Since the source zone of the 2011 Tohoku Tsunami was closer to land than that of the past Sanriku Tsunamis, the tsunami reached the land so quickly that the people there did not have enough lead time for evacuation. Besides, the magnitude of the earthquake was so large that it took a long time to estimate how big the earthquake was; as a result, the magnitude of the earthquake and anticipated height of the tsunami informed through a tsunami warning was corrected stepwise to higher levels. For example, the warning given immediately after the earthquake anticipated that the tsunami expected to reach the coast of Iwate Prefecture would be 3 m high, but the anticipated height was raised later to 6 m at 3:15 pm and even higher, to 10 m or more, at 3:30 pm Although the information was communicated quickly, we have to admit that many areas suffered power outages and had difficulty in getting the information, which spread the damage.

Some of evacuation sites designated in advance in preparation for earthquakes and tsunamis were flooded. Such places were designated as evacuation places on the basis of past tsunami records. Since conventional disaster-prevention programs did not set guidelines on precautions to take against tsunamis higher than seawalls, the Tohoku Tsunami, which was far taller, eventually did tremendous flood damage.

2.1.2.3 Protection Provided by Seawalls and Its Limitation

Preparations against tsunamis have been implemented comprehensively by combining structural countermeasures, such as seawalls, and nonstructural countermeasures, such as warning and evacuation programs. The scale of the 2011 Tohoku Tsunami was far beyond the design conditions of protection structures in many places along the coast from Hokkaido to Kanto, and the tsunami went over the seawalls there, and ruined many of them completely. This has led many people to point out the limitations of protection by structural measures.

The crown of the seawalls provided along the coast of northern Ibaraki Prefecture to Iwaki City, Fukushima Prefecture was designed to be approximately 5–6 m higher than the mean sea level of Tokyo Bay, according to the design wave height calculated based on the data of wind waves. The 2011 Tohoku Tsunami reached as high as 5–9 m on the coast and went over seawalls into many land areas. Nakoso Coast of Iwaki City, Fukushima Prefecture experienced tsunami damage, but there was an obvious difference in the extent of the damage between the different heights of the seawalls, as shown by Figs. 2.1.4 and 2.1.5 [6]. A survey covering the area from

Fig. 2.1.4 Insignificant damage behind a high seawall in Ooshima District, Nakoso, Fukushima Prefecture. *(Data from Sato S. Seawall performance along southern coast of East Japan impacted by the 2011 Tohoku Tsunami; a note for the reconstruction process. In: Post-tsunami hazard, reconstruction and restoration. Springer; 2014. p. 191–210 [chapter 13]. Mar. 25, 2011. The location is marked in Fig. 2.1.5).*

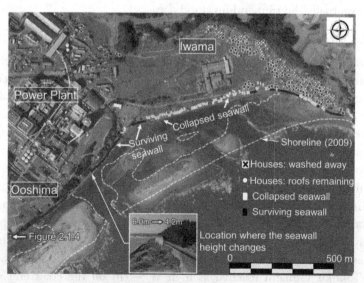

Fig. 2.1.5 Damage due to the 2011 Tohoku Tsunami in Nakoso, Fukushima Prefecture. *(Data from Sato S. Seawall performance along southern coast of East Japan impacted by the 2011 Tohoku Tsunami; a note for the reconstruction process. In: Post-tsunami hazard, reconstruction and restoration. Springer; 2014. p. 191–210 [chapter 13]).*

Chiba Prefecture to Fukushima Prefecture has confirmed that the seawalls that survived the tsunami were effective in reducing the volume of the overflowing tsunami, and suggests that detailed consideration of structures that are resilient enough not to be knocked down completely by overtopping tsunamis will contribute to the enhancement of the resilience of coastal communities. For detailed consideration, it is necessary to analyze various cases in exhaustive detail in order not only to classify how seawalls are broken down, but also to prepare detailed plans of how to make seawalls more resilient against overtopping tsunamis.

2.1.2.4 Tsunamis Invading into Rivers and Causing Floods From Rivers

The mouths of rivers, toward which tsunamis tend to concentrate, are likely to be vulnerable to tsunami floods. Photo 2.1.1 shows the 2011 Tohoku Tsunami that invaded into the Kido River in Naraha Town, Fukushima Prefecture [7]. The tsunami also ran up many other rivers, including the Otsuchi River and Kesen River in Iwate Prefecture, the Kitakami River, Natori River, and Abukuma River in Miyagi Prefecture, the Same River in Fukushima Prefecture and the Kido River in Kujukuri Town, Chiba Prefecture. The tsunami, which flowed into the courses of these rivers, not only broke levees and flooded areas to the both sides of the levees, but also went over the levees into areas along roads running parallel to the coast and flooded residential houses, and other buildings along the roads.

Photo 2.1.1 Invasion of the 2011 Tohoku Tsunami to the Kido River, Naraha, Fukushima Prefecture. *(Data from Sanuki H, Tajima Y, Yeh H, Sato S. Dynamics of tsunami flooding to river basin. In: Proc. Coastal Dynamics 2013, Bordeaux; 2013).*

In contrast, other rivers, such as the Kozuchi River in Otsuchi Town, Iwate Prefecture and the Shimbori River in Kujukuri Town, Chiba Prefecture have water gates at their mouths, which successfully reduced flood damage. These are important examples demonstrating that the water gates at the river mouths were effective in blocking the flooding tsunami and mitigating flood damage. On the other hand, in the area near the mouth of the Same River in Fukushima Prefecture, many levees and seawalls, particularly connections between levees and seawalls, were heavily damaged. Levees, which are covered with concrete only on the riverside slope, are more vulnerable to overflowing tsunamis than seawalls, which are covered with concrete on three sides: riverside and landside slopes and levee crown. The Same River in Fukushima Prefecture experienced serious floods approximately 1 km upstream from the river mouth and in another point where the smaller Shibu River comes together with the Same River. It was estimated that the tsunami that went up the Same River entered the Shibu River, which has its levee crown at a lower level, and caused a flood there. This case suggests that it is important not to consider preparations against tsunamis in a governmental silo-approach manner, in which, for example, rivers and coasts, or large and small rivers, are respectively overseen by different governmental departments, but to review coasts and rivers, as well as the preparations, from a more general perspective, and from multiple perspectives, to identify structural weakness of preparations against tsunamis and discuss comprehensive countermeasures.

REFERENCES

[1] TTJS (Tohoku Tsunami Joint Survey Group). Tsunami survey database, http://www. coastal.jp/tsunami2011; 2011 [accessed Aug. 2015].
[2] Japan Tsunami Trace Database. http://tsunami-db.irides.tohoku.ac.jp. JNES and Tohoku University; 2015 [accessed Aug. 2015].
[3] Tsunami Joint Survey Group. Photo archive. http://grene-city.csis.u-tokyo.ac.jp; 2013 [accessed Aug. 2015].
[4] Hori M, Satake K, editors. The sciences of the Great East Japan Earthquake. Tokyo: University of Tokyo Press; 2012 p. 272 [in Japanese].
[5] Sato S. Characteristics of the 2011 Tohoku Tsunami and introduction of two level tsunamis for tsunami disaster mitigation. Proc Jpn Acad Ser B 2015;91:262–72.
[6] Sato S. Seawall performance along southern coast of East Japan impacted by the 2011 Tohoku Tsunami; a note for the reconstruction process. In: Post-tsunami hazard, reconstruction and restoration. Springer; 2014. p. 191–210 [chapter 13].
[7] Sanuki H, Tajima Y, Yeh H, Sato S. Dynamics of tsunami flooding to river basin. In: Proc. Coastal Dynamics 2013, Bordeaux; 2013.

CHAPTER 2.2

Super Typhoon Haiyan in the Philippines

Y. Tajima, T. Shimozono
Department of Civil Engineering, The University of Tokyo, Tokyo, Japan

2.2.1 WHY SERIOUS DAMAGE WAS CAUSED IN THE PHILIPPINES BY SUPER TYPHOON HAIYAN?

In order to improve resilience in areas heavily damaged by a great flood, it is important to grasp the hydrological characteristics of the flooded area as precisely possible, estimate various risks including multiple causes of disasters, and adopt practical measure for reducing the causes one by one.

The 30th typhoon (international name: Haiyan) that developed in the Northwestern Pacific Ocean in Nov. 2013 intensified as it tracked north northwest, reached 895 hPa at 21:00 Nov. 7, half a day before it hit the Philippines, and reached Samar Island and Leyte Island in the early morning of Nov. 8 with a maximum sustained wind of 90 m/s. According to Weather Underground [1], Haiyan was the fourth largest typhoon in scale and the largest tropical depression in recorded history for maximum sustained wind when it reached the Philippines. The strong wind caused by Haiyan, storm surge and storm waves caused by the strong wind affected more than 16 million people, and caused 7354 fatalities. About 1.14 million houses were destroyed. Total financial damage to infrastructure, fisheries, and agriculture was more than 39.8 billion pesos (as of Apr. 2014, National Disaster Risk Reduction and Management Council [2]).

Various geographical factors and social factors are considered to have contributed to the tremendous damage of this largest typhoon. Here, we shall summarize the results of the on-site investigation that was jointly implemented by the Japan Society of Civil Engineering (JSCE) and the Philippines Institute of Civil Engineering (PICE) [8] and add consideration of how to improve of resilience.

2.2.2 OUTLINE OF THE STORM SURGE DAMAGE IN THE PHILIPPINES CAUSED BY TYPHOON HAIYAN

2.2.2.1 Inundation Height Distribution

The path of Typhoon Haiyan is shown in Fig. 2.2.1, and inundation height distribution obtained by the investigation is shown in Fig. 2.2.2. Since many local residents stayed in their homes in the coastal area and witnessed the flood, inundation heights and upstream heights shown in the figures are mainly based on the stories of such residents (place and inundation height they experienced). As is shown in Fig. 2.2.2, a high inundation height is recorded in the deep San Pedro Bay, and a high tsunami trace height was observed on the coast west of San Pedro Bay (east coast of Leyte Island) even in the area near the northern mouth of San Pedro Bay, immediately north of the path of Haiyan. Furthermore, inundation height and tsunami trace height observed on the east coast of Eastern Samar were higher than those observed in the deep San Pedro Bay.

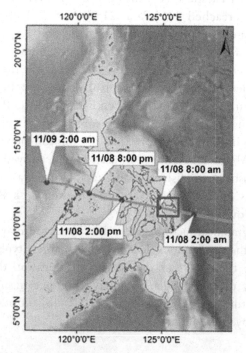

Fig. 2.2.1 Track of Typhoon Haiyan [3].

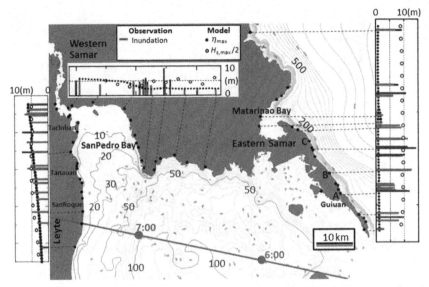

Fig. 2.2.2 Alongshore distribution of measured and computed inundation heights of storm surges and storm waves due to Typhoon Haiyan [3].

The damage by area is as follows:

(i) West mouth of San Pedro Bay

At many places on the coast west of the mouth of San Pedro Bay, inundation height was as high as that in the deep bay where usually highest inundation height is observed in case of storm surges. This implies that not only storm surges but also storm waves intensified the damage. For example, it was witnessed that inundation height increased gradually due to multiple storm waves at places A, B, and C (Fig. 2.2.2) located near the place where Haiyan landed. In addition, three strong waves hit place D, where a strong stream from northeast to southwest was observed in the flooded area. Local residents who were knocked over and carried away by the flood swam to evacuate.

(ii) Deep in San Pedro Bay

In deep San Pedro Bay, eyewitness information on damage has different characteristics from the information obtained in the mouth of the bay. The water level in the deep bay reached its peak more than 2 hours after the water level reached its peak at the mouth of the bay, and maintained its peak for about an hour. While remarkable flooding due to storm surge continued for a long time, a sharp change of water level due to waves was observed at place E in the coastal area. The basements of many concrete houses were scoured out violently. Pillars and joists of the houses were destroyed (Photo 2.2.1).

Photo 2.2.1 Collapsed concrete buildings at Barangay 87 (location E in Fig. 2.2.2).

(iii) East mouth of San Pedro Bay

On the east coast of San Pedro Bay, inundation height gradually decreased from deep in the bay to the mouth of the bay. A sharp change of water level due to waves was observed at place G where the coastline stretches to the southwest. On the other hand, the water level increased gradually and flooded place F that is surrounded by an indented coastline and faces northwest.

(iv) East coast of Eastern Samar

The east coast of Eastern Samar is fringed with 200–700 wide coral reefs, and the Philippine Trench lies offshore. Therefore, elevation of the water surface due to drift was not very significant. On the contrary, it is assumed that the coast was affected greatly by storm waves because it faces the Pacific Ocean.

In the settlement located in the coastal area of east of Salcedo (location H), water advanced as far as a coconut forest at an altitude of about 10 m and about 200 m from the coastline. We also heard oral evidence of a resident who was in a house in the coastal area that was engulfed in violently rolling waves many times. When the typhoon hit the area the coastline was pushed more than 50 m inland and the coconut forest was strongly eroded along the sea side. The seashore around coconut trees was eroded by a maximum of 2 m and coconut roots were exposed (Photo 2.2.2). While most of the coastline of Eastern Samar including Salcedo is fringed with a 100–200 m wide coral reef, the area near location I, which is fringed with coral reef with a width of about 750 m and has the vast low-lying area. The water reached a point more than 1400 m away from the coastline.

Photo 2.2.2 Exposed roots of pine trees due to erosion of sandy beach [4].

2.2.3 ALONGSHORE TOPOGRAPHY AND RESILIENCE AGAINST STORM SURGE AND WAVES

It was revealed that not only the height of the water level in the inundated area but also the time when inundation started, duration of peak water level, and fluctuation of water level were significantly different according to the place in Leyte Island and Samar Island. This section investigates the hydrodynamic characteristics of storm surges and waves based on numerical analysis and describes influence of nearshore topography on regional resilience.

2.2.3.1 Numerical Investigations of Characteristics of Storm Surge and Waves

Storm surge is explained by the two mechanism of water level increase: water level increase due to "pressure surge," which is caused when the low pressure typhoon passes over the sea, and water level increase due to "wind-driven surge," which is caused by the strong wind blown over the sea. The numerical model used here is to estimate the wind conditions based on the typhoon path and pressure distribution, and we use such estimation as input conditions to duplicate the storm surge caused by "pressure surge" and "wind-driven surge." The height of wind-driven surge tends to be greater in a shallow sea because the surge-induced hydrostatic pressure force acting against the wind-blowing direction tends to be smaller. It is therefore known that water level of shallow and closed bay tends to increase by drift.

On the other hand, violent fluctuation of water level caused by strong surface wind generates wind waves. Wave height tends to increase by the wind speed or the distance traveled by wind (fetch). A wind wave model is to duplicate the phenomenon in which the wave develops and propagates under strong wind. While the wind-driven surge tends to amplify in a shallow sea, wave intensification does not depend on the water depth. Rather, waves tend to propagate instead of damping when water is deep, because the influence of frictional resistance from the seafloor is weakened.

Calculation of storm surge and storm wave by the model is shown with the investigation result of Fig. 2.2.2 (bar graph). The black dots in the figure show the distribution of the highest surge height to the coastal direction, and outlined circles show the most significant wave height based on the calculation result of waves. Calculation of peak water level based on the calculation result of storm surge of the west coast of San Pedro Bay monotonously and exponentially increases from the bay entrance to deep in the bay. Inundation depth obtained from on-site investigation shows rise and fall regionally and also shows that inundation depth is relatively high at the bay entrance. On the other hand, calculation result of the wave (solid white circle) shows the trend that the wave becomes greater at the bay entrance and smaller deep in the bay, which implies that flood in the bay entrance area of San Pedro Bay was affected not only by storm surge but also by storm waves.

Next, when focusing on the east coast of Eastern Samar, Fig. 2.2.2 reveals that inundation depth measured at the disaster site significantly exceeds the calculation result of reproduced storm surge. On the contrary, the calculated storm wave is extremely large and it is estimated that storm wave was the biggest cause of disaster in Eastern Samar. The width of the coastline in Eastern Samar is from 100 to 800 m, as shown in Fig. 2.2.3, surrounded by coral reefs and so shallow that some parts of coral reefs are exposed above sea level at low tide. Waves normally break at the outer rim of coral reefs and so lapping waves are relatively quiet. According to local residents, they have never experienced such a flood that reached their residential area. This is considered one of the reasons that not a few residents failed to evacuate. Shimozono et al. [5] performed numerical simulation under the conditions that a storm wave caused by strong typhoon hits. The result shows that the water level rises abnormally under certain conditions when an extremely big wave attacks. As identified above, since even a coast "protected" by coral reefs may be flooded depending on the conditions, it is important to evaluate the hazard of each region appropriately, based on various assumptions, and to inform local residents of these hazards in an understandable way.

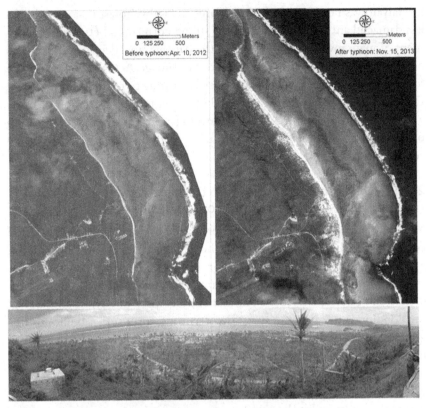

Fig. 2.2.3 Coral reef at Guiuan (location I in Fig. 2.2.2) [4].

2.2.3.2 Development of Mangrove Forest and Resilience of the Region

Mangrove often attracts interest as a disaster mitigation measure, especially in developing countries. In Matarinao Bay, which is located on the east coast of Eastern Samar, mangrove forest grows well, which reduced damage dramatically even in some areas near the bay entrance although its geographical formation widely opens to the direction that storm surge hit [4]. In addition, a high level of correlation was identified between locally mitigated damage and development of mangrove forest before the disaster. Based on the research of Gunasekara et al. [6], the main reason for a suddenly damped wave is the breakage and refraction of the wave by sudden change of water depth, and this is the phenomenon typically identified on coasts where coral reef is developed. While the small mangrove forest that grew in the devastated area was seriously damaged by the storm wave, significant damage due to Haiyan was not identified for the large mangrove forest that grew in the less damaged area. In Matarinao Bay, a strong correlation between development of mangrove forest and scale of the damage was identified. But it was not the mangrove forest that mitigated the damage. Actually, the mangrove forest grew in the area where waves were small due to geographical features. Consequently, it is considered not only that aggressive forestation of mangrove is effective for improving disaster mitigation, but also that monitoring of the forestation status works as an index for resilience assessment of the coastal area relying on the natural conditions.

2.2.3.3 Difference of Inundation Damage by Geographical Features, Consideration of Improving Resilience

Mori et al. [7] calculated several patterns of storm surges by changing the conditions, including the path of typhoon, pressure pattern, and typhoon eye, and pointed out that the elevation of the sea level deep in San Pedro Bay due to storm surge was significantly affected by the above typhoon conditions. They also pointed out that Haiyan was a rare typhoon that had the conditions to cause the worst disaster from storm surge. Typhoons frequently hit the Philippines. Local residents said that they had experienced huge typhoons, but they had never experienced such enormous inundation damage. This is considered to be one of the reasons why many residents carelessly did not evacuate, even though a warning had been issued a day before the typhoon hit the area.

In the bay entrance area, not a few residents were caught by the waves in an area ordered to be evacuated because not only storm surge but also storm waves hit the area and elevated the sea level several times. Furthermore, it was revealed that the damage was increased because a never-before-experienced elevation of the sea level occurred on the coral reefs which usually protect the coast. It is assumed that large deviation between the risk that the local residents estimated and the actual disaster is one of the factors that intensified the loss of life.

REFERENCES

[1] Masters J. Jeff Masters Blog, Weather Underground, http://www.wunderground.com/blog/JeffMasters/archive.html?year=2013&month=11&MR=1 [accessed in 2014].
[2] National Disaster Risk Reduction and Management Council. NDRRMC update, SiteRep no. 108 effects of Typhoon "Yolanda" (Haiyan); 2014.
[3] Tajima Y, Gunasekara K, Shimozono T, Cruz EC. Study on locally varying inundation characteristics induced by super Typhoon Haiyan. Part 1: dynamic behavior of storm surge and waves around San Pedro Bay. Coast Eng J 2016;58(1).

[4] Tajima Y, Shimozono T, Gunasekara K, Cruz EC. Study on locally varying inundation characteristics induced by super Typhoon Haiyan. Part 2: deformation of storm waves on the beach with fringing reef along the east coast of Eastern Samar. Coast Eng J 2016; 58(1).

[5] Shimozono T, Tajima Y, Kennedy AB, Nobuoka H, Sasaki J, Sato S. Combined infragravity wave and sea-swell runup over fringing reefs by super Typhoon Haiyan. J Geophys Res Oceans 2015;120(6):4463–86. http://dx.doi.org/10.1002/2015JC010760.

[6] Gunasekara K, Tajima Y, Shimozono T. Variation of impact along the east coast of Eastern Samar due to Typhoon Haiyan in the Philippines. J JSCE, B2 (Coast Eng) 2014;70(2):I_241–5.

[7] Mori N, Kato M, Kim S, Mase H, Shibutani Y, Takemi T, et al. Local amplification of storm surge by super Typhoon Haiyan in Leyte Gulf. Geophys Res Lett 2014;41(14):5106–13. http://dx.doi.org/10.1002/2014GL060689.

[8] Tajima Y, Yasuda T, Pacheco BM, Cruz EC, Kawasaki K, Nobuoka H, et al. Initial report of JSCE-PICE Joint Survey on storm surge disaster caused by Typhoon Haiyan. Coast Eng J 2014;56:1450006. http://dx.doi.org/10.1142/S0578563414500065.

CHAPTER 2.3

Damage From and Resilience Against the 2011 Thai Flood

S. Nakamura
Department of Civil Engineering, Nagoya University, Nagoya, Japan

2.3.1 IMPACTS OF 2011 THAI FLOOD

In 2011, when the Great East Japan Earthquake occurred, the total world-wide natural disaster damage for the year reached $366 billion—the largest since 1990 [1]. Nearly half of the damage of that year was caused by the Great East Japan Earthquake, and almost all the remaining was related to water disasters. Above all, in Thailand, a flood occurred from Oct. to Dec. 2011. The Chao Phraya River that runs from north to south through the central part of the country overflowed its levees and flooded the capital city of Bangkok. Not only houses but also many foreign-capital plants operating in Thailand were flooded. The economic damage exceeded $40 billion. The economic damage was enormous and drove down the country's real GDP growth rate for the year to 0.07% from an initially anticipated 8% [2].

Not only the government but also local residents took many damage reduction measures against this historically massive flood. Although many of these measures did not work sufficiently, this experience gives some hints on how to react to an unexpected massive flood and also how to improve the resilience of the country. In this chapter, based on our on-site investigation conducted under "Integrated Study Project on Hydro–Meteorological Prediction and Adaptation to Climate Change in Thailand (IMPAC-T)" (principal investigator: Taikan Oki, The University of Tokyo) supported by the Science and Technology Research Partnership for Sustainable Development (SATREPS), JST/JICA, Japan. I shall describe the cause and effect of the 2011 Thai flood and the measures actually taken. Then, I shall consider the problems regarding the improvement of resilience.

2.3.2 WHAT CAUSED THE 2011 THAI FLOOD

The Chao Phraya River, which flooded in 2011, is the largest river in Thailand, with a river basin of about 160,000 km^2 (Fig. 2.3.1). The Ping,

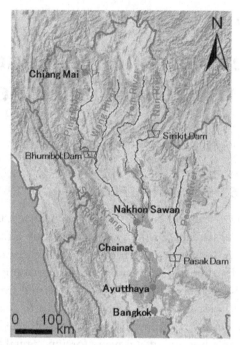

Fig. 2.3.1 Chao Phraya River basin and dams.

Wang, Yom, and Nan rivers join to make the Chao Phraya River in Nakhon Sawan Province, located in the middle basin of the Chao Phraya River area. The Chao Phraya River then goes through the Central Plain and flows into the Gulf of Thailand. The distance from Nakhon Sawan to the river mouth is about 400 km and the stream average gradient is about 1/10,000. The Chao Phraya River is a very low-pitched river. Bhumibol Dam (13.5 billion m³) is located on the upper Ping River and Sirikit Dam (9.5 billion m³) is located on the upper Nan River. Both generate electricity and adjust service water capacity [3].

Most of Thailand has a tropical savanna climate. The rainy season (from May to Oct.) and the dry season (from Nov. to Apr.) are clearly different. In Bangkok, about 90% of annual rainfall falls in the rainy season. The average annual rainfall of the Chao Phraya River area is about 1120 mm, but it reached 1510 mm in 2011—the highest in the past 50 years [3]. There is no doubt that the direct cause of the 2011 Thai flood is this historic rainfall. As a consequence, total discharge of the rainy season of 2011 reached 32.6 billion m³ in Nakhon Sawan, which is located in the downstream area. It was the highest discharge since 1956 and reached 232% of the average discharge [3].

In the delta area from Nakhon Sawan to Bangkok, the gradient of the Chao Phraya River is quite low and the river contains a lot of silt. Therefore, the river winds repeatedly and has an irregular geometry. Fig. 2.3.2 shows the discharge capacity of the Chao Phraya River, which gradually decreases to Bang Sai, the northern area of Bangkok. That is to say, the flood flows coming down from upstream must be diffused into the floodplain. The Royal Irrigation Department constructed levees on both sides of the Chao Phraya River from Chainat through Ayutthaya to protect the rice fields from small and medium floods. Floodgates are used to lead a large flood into the rice fields that are connected to the floodplain through a water channel. This is to prevent an unexpected dike break, which would cause serious damage, and to control a flood flowing into the Bangkok Metropolitan Area from spreading downstream of Ayutthaya. The rice fields are used as a buffer for flood control that is relatively unaffected by flooding. In other words, Ayutthaya and the upstream area work as a retarding basin to protect the Bangkok Metropolitan Area. This retarding basin is called "Monkey Cheek" and some rice fields are designated as an area in which inundation is allowed according to administrative policy. Highly sophisticated judgment is required for diverting the flood to Monkey Cheek, because the judgment

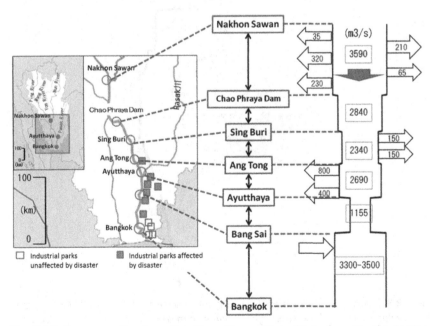

Fig. 2.3.2 Flow capacity along the Chao Phraya River *(Data from Royal Irrigation Department, Thailand)*.

must be made in consideration of the timing of harvesting and the growth situation of each area. In the 2011 Thai flood, a large area of the right levee was inundated in mid–September in addition to the flood flow being more than expected, which broke the levees and floodgates at several locations. The flood flow concentrated on the left bank went downstream for several weeks to hit the historic city of Ayutthaya and reached the Bangkok Metropolitan Area in late Oct. [3].

2.3.3 DAMAGE IN BANGKOK AND RESPONSE OF THE ADMINISTRATION IN THE 2011 THAI FLOOD

The Bangkok Metropolitan Administration drew up a flood control plan in 1999 with the technical support of JICA (Japan International Cooperation Agency) (Fig. 2.3.3). In this plan, the outer dike, "King's Dike," surrounding the Bangkok Metropolitan Administration was to be enhanced. The land surrounding King's Dike was to be designated as a green belt and for flood control function. A diversion channel was to be built from the green belt to

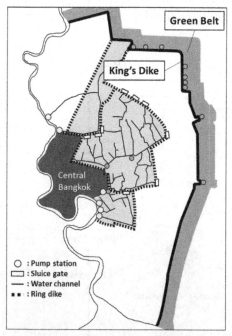

Fig. 2.3.3 Flood prevention plan in East Bangkok. *(Data from Yoshikawa KY, Motonaga Y. Post-evaluative study of flood damage mitigation in the basin of gentle flowing rivers on a low-lying plain. J Jpn Soc Hydrol Water Resour 2006;19(4):267–70 [in Japanese]).*

the Gulf of Thailand [4]. However, due to rapid urban sprawl, a residential area and industrial estates was built on the green belt and the flood control function of the green belt did not work. In addition, Suvarnabhumi International Airport was built on the planned diversion channel route, which derailed the plan of building the diversion channel. Floods flowing into the central area of Bangkok would have been prevented if at least the King's Dike had worked appropriately, but actually, the industrial estates and residential area developed on the green belt were flooded and suffered significant damage. Among the industrial estates, those developed through after the 1980s were built on the hilly area including Laem Chabang that is not suitable for rice cropping. However, industrial estates were developed on the floodplain in and after the 1980s to take advantage of its abundant labor force and vast expense of land. The industrial estate, which requires a broad stretch of land, could only be built on underutilized land with a low safety. Houses were also developed in the neighboring area for the industrial estates and its workers. Such land development without considering flood risk caused many houses and industrial estates to be submerged, and resulted in tremendous damage. Furthermore, this improper land use caused conflict between the residents facing each other across King's Dike, and emergency levees made of sand bags and constructed as a tentative measure against the Thai flood significantly affected the flood prevention activities.

The Chao Phraya River flows slowly, so its flooding speed is very slow compared with that in the Japanese archipelago. Sufficient lead time is available before the arrival of a flood. In the 2011 Thai flood, levees broke and floodgates were damaged in the upper stream of Ayutthaya in early Sep., releasing an enormous flood. In late Oct., the flood reached the Bangkok Metropolitan Region (BMR), located about 100 km down the river. That means there was a lead time of more than a month from dike breakage to the arrival of flood. The Thai government tried to utilize this lead time and built an emergency dike (Fig. 2.3.4) along with the irrigation ditch that stretched from east to west to direct the water coming from the north to the Bang Pakong River through the ditch and by pumping. But this attempt ended in failure, because the flooding was greater than expected and pump drainage to the Bang Pakong River could not keep up with the amount of flooding. Also, there was trouble in building and maintaining emergency levees due to conflict among the residents across the emergency levees, and opposition campaigns frequently occurred. Residents launched such opposition campaigns in the whole watershed area in the time of the 2011 Thai flood, which caused trouble in smooth flood control.

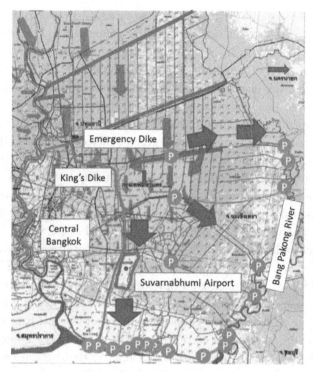

Fig. 2.3.4 Flood-fighting activities in 2011 flood. P in the figure means pomp station.

In the watershed of the Chao Phraya River, the plan was to set several flood control areas and maintain the reliability of flood control, but actually, due to new land development and the change of lifestyle in such areas, the flood control area was reduced and the reliability of flood control dropped sharply.

2.3.4 IMPACT OF THE 2011 FLOOD ON DAILY LIFE

In the 2011 Thai flood, not only the government but also residents prepared for the flood. Our study group conducted interviews from Oct. to Dec. with the residents of the BMR (a total of 142 interviewees) about their response to the flood. The most common prevention measure was to build a waterproof barrier. In all areas of Bangkok, it was often seen that residents built waterproof barriers with sand bags or concrete blocks (Photo 2.3.1). The second major preventive measure was to carry the household goods upstairs. Considering that the flood height did not reach the second floor, it was an effective measure. Stilted houses were traditionally built in Thailand as a flood prevention measure. But since the lifestyle has changed (to Western style) recently, the main living space has shifted to the first floor. This might be one reason that the damage was intensified and also a reason for carrying

Photo 2.3.1 Flood walls in Bangkok, 2011 flood.

up the household goods. Storage of food and drinking water, and installation of a pump, were other preventive measures. All such measures were carried out in time because there was enough lead time and expectation of damage.

On the other hand, a phenomenon occurred that was beyond the scope of anticipation. This was the prolonged immersion period which was characteristic of a delta region. For example, an industrial estate located north of Bangkok was immersed from Oct. 17 and the flood water was not completely gone until Dec. 8. The flood continued for nearly 2 months. The residents of the affected area had to live in flooded conditions. The transportation network was cut for a long time, which severely affected daily life. Fig. 2.3.5 shows the largest flooded area (on Nov. 5) in and around the Bangkok Metropolitan Area. Most of the northern part of the Bangkok Metropolitan Area was flooded, and roads, including major highways, were closed. Some residents escaped to evacuation areas (42%, according to our survey). More than 50%

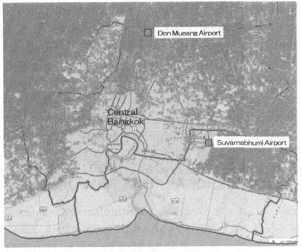

Fig. 2.3.5 Flood inundation area around Bangkok on Nov. 15, 2011.

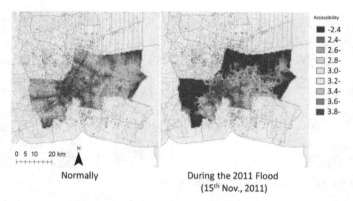

Fig. 2.3.6 Accessibility of urban facilities in Bangkok. *(Data from Yamashita Y, Nakamura S, Sugimoto K, Hayashi Y, Vichiensan V, Kato H. Assessing flood resilience of transport networks by measuring damages to living opportunity—case of Thai flood 2011. World conference on transport research—WCTR, Shanghai; submitted for publication).*

of residents stayed on the second floor of their houses and used small boats to buy food and household things, go to hospital, and commute. They gave the following reasons: "We need to keep watch for theft," "The evacuation place was already full," "We don't feel inconvenience because electricity, water, and food are available at home," and "Flooding is an annual event and we are accustomed to it." In the 2011 Thai flood, many residents stayed in their homes. Also, their opportunities in daily life were dramatically decreased because the transportation network was shut off by flooding.

Fig. 2.3.6 shows the total accessibility of hospitals, shops, and schools in the Bangkok Metropolitan Area in normal times and during the 2011 Thai flood (as of Nov. 2015, when the flooded area was largest). Accessibility in normal times is high in the central area of Bangkok and along the highway. On the other hand, when looking at accessibility during the flood, although it is high in the central area, which was not affected by flooding, accessibility decreased dramatically in the north, east, and west areas where highways were flooded. These areas are developed into industrial estates and residential areas. Daily life in the newly developed areas was affected the most. In order to improve the resilience of these areas, it is important to secure the transportation network in case of flooding as well as preventing the facilities from flooding.

2.3.5 LESSONS LEARNED FROM THE 2011 THAI FLOOD TO IMPROVE RESILIENCE AGAINST FLOODING

The northern part of the Bangkok Metropolitan Administration and its surroundings which were heavily affected by the 2011 Thai flood have developed along with the growth of the Thai economy. Many of the flooded industrial estates were established in and after the 1980s, and were the driving force of economic growth in Thailand after the 1990s. Unfortunately, since the global supply chain was shut down due to the flood, the economic impact spread, not only in Thailand, but also to all over the world. Flooding used to be regarded as a local issue, but it may now become a global issue, depending on the disaster location. Improvement of resilience against flooding is not just a local issue, but becomes more and more global.

On the other hand, the natural phenomenon of flooding is closely connected with the global issue of climate change. Recent climate change research predicts increasing flood risk in the Asian monsoon region, which includes Thailand [5]. Adaptation for climate change is an important issue. Above all, how to include the external factor of climate change into conventional flood control measures is an issue not only for developing countries but also for developed countries. Efforts to establish practical adaptation measures have just begun in some countries.

Previously, levees and dams were the major flood control method. However, in order to improve the resilience of urban areas against accelerating climate change, restructuring of urban areas is necessary in addition, to consolidate these existing flood prevention facilities. Withdrawal from areas with high risk of flooding and consolidation to areas with low risk of flooding, and enhancement of the transportation network connecting the consolidated areas, will realize urban efficiency, a more comfortable life, and also better resilience against natural disasters. That is to say, consolidation of flood prevention facilities and restructuring of the urban area are inseparable. Furthermore, it is considered that lifestyle change worsened the flood damage. Re-evaluation of traditional stilt houses of Thailand that fit the climate and the renovation of these traditional houses are also considered an important issue. In order to prevent similar flooding from occurring, discussion and action to improve resilience against flood are desired.

REFERENCES

[1] UNEP. Annual report 2011, United Nations Environment Programme. 2012.
[2] The World Bank. Thai flood 2011: rapid assessment for resilient recovery and reconstruction planning. World Bank. 2012.
[3] Komori D, Nakamura S, Kiguchi M, Nishijima A, Yamazaki D, Suzuki S, et al. Characteristics of the 2011 Chao Phraya River flood in Central Thailand. Hydrol Res Lett 2012;6:41–6.
[4] Yoshikawa K, Motonaga Y. Post-evaluative study of flood damage mitigation in the basin of gentle flowing rivers on a low-lying plain. J Jpn Soc Hydrol Water Resour 2006;19(4):267–70 [in Japanese].
[5] Hirabayashi Y, Mahendran R, Koirala S, Konoshima L, Yamazaki D, Watanabe S, et al. Global flood risk under climate change. Nat Clim Chang 2013;3:816–21.

CHAPTER 2.4

Disaster Resilience Learned from the 2007 Earthquake in Peru

T. Inamura
The Open University of Japan, Chiba, Japan

2.4.1 AN EARTHQUAKE THAT HIT PERU

A big earthquake of magnitude 8 hit Peru on Aug. 15, 2007. The epicenter was located just offshore of Pisco, a town about 300 km to the south of the capital city of Lima, with a flourishing fishery. The date when the earthquake hit was also the opening day of a special exhibition at Museo Amano (an archaeological museum in Lima) regarding the "Ruins Las Shicras" that were found in the Valley of Chancay. The history of the Ruins Las Shicras can be traced back to around 3000 BC. I participated in the archaeological excavation that Museo Amano organized. One of the excavation themes was about structures to cope with earthquakes.

A few days after the earthquake struck, I visited Pisco with one of my Peruvian friends, who had reported on the earthquake as a newscaster. Most of the houses built of sun-dried adobe bricks had collapsed. More than 100 people were victims of the collapse of a Catholic church in the main square (Photo 2.4.1). The earthquake also caused a tsunami, which washed most of the fishery boats ashore and devastated the fishing industry.

As stores had been looted right after the earthquake, armed vehicles were strictly policing in the town (Photo 2.4.2). Relief supplies were brought to Pisco by the navy, which has a base in the neighboring port (Photo 2.4.3). The soldiers kept the disaster victims in line and handed them relief supplies.

As a researcher of cultural anthropology, I have been studying the community of indigenous people living in the mountainous region of Peru since 1978. I have encountered the wisdom of the descendants of the Inca, people who adapted themselves to the harsh natural environment of the high mountainous region and made good and sustainable use of the wide variety of resources. I was also interested in the "wisdom of resilience" of ancient civilizations found through the excavation of the Ruins Las Shicras and other such like ruins. I saw the reality of disaster with my own eyes and I keenly felt the artificial aspect and cultural aspect of disaster, through

Photo 2.4.1 Catholic church destroyed by the earthquake.

Photo 2.4.2 Peacekeeping operations by the army after the earthquake.

Photo 2.4.3 Relief supplies delivered by the navy.

comparison with the resilience of ancient civilizations, the sustainability of the communities of indigenous people, and the vulnerability of the contemporary urban communities, and also from the contrast of these three factors. That is to say, even though the disaster can be attributed to a natural phenomenon, the development process of the disaster and the recovery from that disaster are social and cultural phenomena. In 2008, when a severe earthquake occurred in Sichuan, China, I visited Chang and Tibetan communities to do aid work. Furthermore, I witnessed the disaster of the Great East Japan Earthquake in 2011. I became increasingly convinced of these aspects through my experience.

Natural disasters are categorized into three areas: tectonics-related disasters, including earthquakes, tsunamis, and volcanic eruptions; air-related disasters, including drought, damage due to cold weather, severe rainstorms, and lightning strikes; and phenomena occurring on the Earth's surface, including landslides, flooding, and fires. Although these disasters are natural phenomena, the impact of human activities on the atmosphere and the Earth's surface is apparent. Greenhouse gas emissions cause global warming, which in turn causes abnormal weather, triggers natural disasters, and deteriorates agricultural productivity. Deforestation causes landslides and downstream flooding.

The impact of human activities on disasters is far greater than that of the causes I have mentioned above. The extent of earthquake damage varies greatly according to the structure of a city or its buildings, site locations, and habitation patterns. The magnitude of damage is influenced by social disparities within the local community. Poorer people and socially vulnerable people are more likely to be endangered. Social structure is also greatly connected with disasters. While dictatorship, social disparities, corruption, etc. are major factors in exacerbating the effects of the damage, traditional mutual aid systems work to reduce damage and promote recovery. Epidemics right after a disaster, provision of relief to the affected area, the recovery plan, and impacts of disasters on the society differ according to community. In a disaster, the natural environment and social structure mutually affect each other.

2.4.2 RESILIENCE OF ANCIENT ANDEAN CIVILIZATION

It was a tradition of the Andean civilization to cover an old temple with a new temple building. For the pyramid type temples of 3000 BC, fillers called *shicra*, which were stones wrapped with netting made of plants, were used to cover the old temple. The excavation of the ruins in the Valley of Chancay began in order to investigate a previous hypothesis that *shicra* had

Photo 2.4.4 Sun god temple in the ruins of Machu Picchu.

antiseismic capabilities similar to that of sandbags. It is difficult to verify if *shicra* actually were intended to have antiseismic capabilities, but multiple researchers have pointed out that the buildings of Andes civilization are constructed to withstand earthquakes.

The last civilization of the ancient Andes regions, the Inca civilization (AD 1420–1533), is famous for its huge buildings made of stone. Inca people used their ingenuity in designing such buildings to endure heavy loads and natural disasters, including earthquakes. The rock walls of the Inca ruins such as Machu Picchu are slightly inclined toward the inside of the building, windows and entrances are designed to be trapezoidal (Photo 2.4.4), and rocks are especially embedded in the wall corners. The huge retaining walls of Sacsayhuaman ruins have a zigzag shape, which is considered to be extremely resistant to earthquake waves from any direction.

Pyramids found in the Cahuachi ruins of the Nazca civilization (AD 100–600) are made of sun-dried bricks. Sand and plants make a layer in some parts of its inner structure. Orefichi, an archaeologist who excavated the ruins, pointed out that the structure has a kind of earthquake-resistant effect as well as a load reduction effect [1].

The Andes Mountains, which were formed at the boundary of two tectonic plates, is an earthquake-prone region. In addition, the El Niño phenomenon causes natural disasters including floods and landslides, due to heavy rain in the coastal area, which usually has little rain. There is no doubt that ingenuity in disaster reduction was developed for the huge buildings in the Andes from the ancient age in response to natural disasters.

2.4.3 SUSTAINABILITY IN THE COMMUNITIES OF INDIGENOUS PEOPLE

As the Central Andes are high mountains located in a tropical area, there is a wide variety of nature depending on the altitude. The ancient Andean civilization utilized the variety of nature and developed based on the cultivation of various plants, including potatoes and corn. Even now, many indigenous people inherit a part of the tradition of the Inca era and maintain sustainable lives in the highlands of Peru.

People keep llamas and alpacas in the highlands, at more than 4000 m elevation in the southern part of Peru. Farmers live in the steep valley region at an altitude below 4000 m. Pastoralists keeping llamas and alpacas do not take milk from their livestock but live on crops. They obtain the crops through the barter of meat from their livestock and rock salt with farmers, or by using llamas to transport crops from terraced fields to the farmers' houses (Photo 2.4.5). The pastoralists of the Andes live through close ties with farmers [2].

Farmers cultivate various agricultural products according to the altitude. Diversity not only of the crop types but also of the varieties is important for agriculture in the Andes. For example, around 20–30 varieties of potatoes are cultivated in one village and each variety has a different folk term. This diversity works to disperse the risks of climate change, disease and insect damage. Conventional agriculture maintains a combination of fallow period and crop rotation in a cycle of 5–6 years.

Mutual assistance among community members is also developed. This is, for example, barter of labor, such as planting and harvesting in the agricultural community and shearing of alpacas in the community of pastoralists.

Photo 2.4.5 Caravan of llamas on the Andean plateau.

Other events that sustain the ties of the community are cooperation in repair of irrigation system and roads, maintenance of churches, and shared beliefs and feasts in festivals.

2.4.4 PISCO, FOUR YEARS LATER

I visited the disaster site, Pisco, again in mid-August of 2011. A new church was under construction in the central square, where the former one had collapsed. The municipal building was left collapsed the same as it had been right after the earthquake. Temporary houses were built on the coastal street (Photo 2.4.6). When I visited one of the houses, the owner showed me his house. The frontage was about 4 m, there was a kitchen inside the room of about 4 m depth which was used as a living room and also as a bedroom. Outside the house was the beach. The house was for a six-member family. The head of the family, Pablo, his wife, and his daughter, who was around the age of 20, talked about their life after the earthquake.

They said, "We were about to take supper. We felt an enormous shaking and everything collapsed. So, we evacuated to the park in the neighborhood. Afflicted people gathered in the park. There was no food or water that night. It was scary, but nobody knew where to go. The next day, we went home to find that our house had collapsed due to the tsunami. So we looked for blankets, etc. and went back to the park. People of the community brought rice, beans, and fish, cooked at a temporary communal kitchen, and ate together."

Photo 2.4.6 A family who have been living in temporary housing for 4 years since the tsunami.

They also said, "A government truck distributed relief supplies and then drove off. We ran after the truck and scrambled for the supplies. We ran after the truck just like dogs. We felt so miserable."

"About half a month later, the government provided us with a small camping tent. Then, we built a hut with vinyl and wooden material that Father José of the church in the neighborhood gave us. It was about a year later when temporary houses were built."

I asked them, "How do you make a living?" The head answered, "I am jobless. So, I clean this area and receive some money from Father José." The daughter said, "I want to work. I watched the tsunami that hit Japan on the Internet. Japanese people had a more difficult time than us." The wife said, "We must work harder because we survived." They are very good people, but it seemed to be taking far more time for them to get back to living a self-sustained life.

The main industry of the port town of Pisco is fishing. I visited the fishing port to meet the chairman of the fishermen's cooperative. He said, "The things on the shelf fell down with the huge shaking of earthquake. I went to the breakwater and saw that the water drew back. I told everyone to run away and I also ran. Although both of my feet were caught in the tsunami, the tsunami was not so big and I survived."

Before the earthquake, about 700 fishermen owned 150 fishing boats. The tsunami height was about 2 m, and nearly 80% of the boats were damaged. At first, fishermen could not go fishing due to the trauma, but they gradually resumed fishing two months after the earthquake.

The port building was reconstructed in Jun. 2011. Currently, there are more than 1500 fishermen and about 300 fishing boats (Photo 2.4.7). The number was doubled because the fishery absorbed the jobless people of other industries

Photo 2.4.7 The fishing port was revived after the earthquake and tsunami.

Photo 2.4.8 Small fishing boats that have doubled in number after the revival of the fishing industry following the tsunami.

(Photo 2.4.8). Another big change is that about 50 small cooperative associations were newly established. There was only one cooperative association before the earthquake and only a few fishermen were member. The change is because they realized the necessity of mutual assistance after the earthquake.

What I notice at this fishing port is that diversified industries are quick to recover—that is to say, they are resilient, in the time of reconstruction. Not much investment is required to purchase a new fishing boat, so significant assistance was not necessary for the recovery of the fishing industry. The fishery absorbed the unemployed people from other industries and worked as a safety net. In addition, fishermen learned from the disaster and developed a mutual assistance system by establishing many small cooperative associations.

2.4.5 CORRUPTION OF LOCAL GOVERNMENT AND STRONG DISTRUST OF POLITICS

After having walked along the coast, I made a phone call to a local TV station. Two young reporters came to gather information. They said that the mayor designated the tsunami-hazard area as *zona roja* (red zone) and tried to evacuate the residents from the area, but the residents did not want to leave their land. As a result, the area has not been redeveloped and the residents continue to live in temporary houses made of thin plywood walls.

The role of the municipal government is to support the afflicted people, but the assistance is insufficient. As part of the infrastructural development, brick walls were put up in each block. It is however rumored that this has been more to cover up for the delays in reconstruction than for the residents themselves. The residents call the walls *muro de verguenza* (walls of shame)" and view them as a way to hide the slow pace of reconstruction from the president, visiting senior officials and foreigners who come to assist with the disaster recovery.

Pablo and his family members said that there were a lot of relief supplies, but the city officers took all the good stuff. They also said, "Even though it has been four years since the earthquake, nothing has changed. We have no choice but to live our lives in this temporary housing. We went to receive assistance, but we have not been able to receive it. They continued to say 'tomorrow, or the day after tomorrow' and gave us nothing. Some people received assistance and were able to buy things. We want to build a house but we have received nothing."

I also talked with a woman who works as a fish broker. She now lives in a temporary house with eight family members. She had built a small part of the house in brick. She said, "My house was destroyed in the earthquake and subsequent tsunami. I lost all of my capital. I received one payment of 6000 *soles* (about $2000) of aid money. Since the prices of brick and cement went up enormously after the earthquake, I could only build a basement with the 6000 *soles*. I borrowed some money and build the brick walls a little bit at a time. They say it is *zona roja*—that it is prohibited to build a house—but all I can do is to work and build my house little by little. It's already been four years since the earthquake but nothing has changed. It's as if the earthquake hit yesterday. The government did not provide any earthquake relief. Some people could not receive anything. I have no idea why."

2.4.6 VULNERABILITY OF PERUVIAN COMMUNITIES DURING DISASTERS

The people of the Andes developed resilience against natural disasters before they were conquered by the Spaniards. Although the unity of the wide area of the Andes was lost in the conquest, the indigenous people living in the highlands of the Andes maintain a culture that prioritizes diversity, mutual assistance, and unity in the community. In short, sustainability is maintained in regular vocations and communities there. Sustainability is an accumulation of wisdom to fit the ever-changing natural environment and to maintain daily life. On the other hand, resilience is an accumulation of the wisdom to handle rapid and drastic change. Both were developed in the Andes before the conquest by the Spaniards.

According to Oliver Smith, the forms of adaption to disaster were control of diverse ecosystems, scattered residential patterns, proper housing material and technology, ideology, and an explanatory model [3]. Those were characteristics of the resilience found in the Andes. However, after the Spanish conquest and colonial occupation, the traditional infrastructure and the culture that supported such an infrastructure collapsed. In the society, Europeans and some of the *mestizos* monopolized the state power and indigenous people were pushed to the margins of the society. Even after the independence of Peru in 1821, a few elites and the military forces controlled politics until the 1960s. As a result of latifundism and monoculture, such as production of sugar cane and cotton, indigenous people were forced to live harsh and poverty stricken lives. A significant social divide and permanent political corruption ascribed to colonial occupation has yet to be resolved, even after agricultural reform and democratization.

Buildings and infrastructure that became popular after colonial occupation weakened resilience. In addition, lack of community ties increased the vulnerability of Peruvian society. Above all, the coastal area has a society without traditional community ties where *mestizo* and indigenous people relocated from mountainous regions. Looting, maintenance of security by military forces, illegal trade and unfair distribution of relief supplies, and delay of reconstruction becomes apparent after following natural disasters and are indicative of a modern Peruvian urban community that has a historically negative heritage due to conquest and the long period of colonial occupation, which is common to all of Latin America.

We could understand from the example of Peru that a natural disaster is not only a "natural" phenomenon, but also very much a cultural and social phenomenon. In order to reduce the effects of such disasters, a trustworthy government and a culture and society that supports the idea of mutual assistance are vitally important.

REFERENCES

[1] Orefici G, Drusini A. Nasca: hipótesis y evidenicias de su desarrollo cultural. Centro Italiano Studi e Ricerche Arqueologiche Precolombiane. Brescia BS, Italy; 2003.
[2] Inamura T. The pastoralism in the Andes and the Himalayas. In Global Environmental Research 6-(1). 2002. Tokyo; 85–102. Inamura T. Las características del uso de camélidos en los Andes: El pastoreo y la resurrección del "chacu", la tradición incaica en el Perú. In Desde el exterior: El Perú y sus estudios. Tercer Congreso Internacional de Peruanistas, Nagoya, 2005, Millones, L. y T. Kato (eds.) Fondo Editorial de la Facultad de Ciencias Sociales: Universidad Nacional Mayor de San Marcos, 2006. Lima, Peru; 35–70.
[3] Oliver-Smith A. Peru's five-hundred year earthquake: vulnerability in historical context. In: Oliver-Smith A, Hoffman SM, editors. The angry earth: disaster in anthropological perspective. New York: Routledge; 2008.

CHAPTER 2.5

Loss of Resilience of Mongolian Nomadic Life Due to Urbanization

S. Ishii
Graduate School of Environmental Studies, Nagoya University, Nagoya, Japan

2.5.1 SUSTAINABILITY AND RESILIENCE IN MONGOLIAN NOMADIC SOCIETY

Mongolian nomadic society is full of good ideas on how to live sustainably. They always pay attention to the condition of the grass, and move from one place to another every season before their domestic animals overgraze the area. They do not stick to one place. When they sense the environment is in danger of deteriorating, they try to avoid any damage by moving to another place.

The Mongolians keep five kinds of domestic animals: sheep, goats, cows, horses, and camels. As the grass that an animal likes differs depending on the kind of animal, this does not place an excessive load on the grass plain. Mongolia is often hit by severe droughts and *dzud* (cold-weather damage and snow damage). So, if they depend on only one kind of domestic animal, their risk of losing all of their domestic animals is high. Therefore, they keep a variety of domestic animals in order to avoid such a situation. Nomads are well aware that if resilience against a one-time disaster is insufficient, long-term sustainability put in peril. They have the traditional wisdom always to be prepared for this kind of situation.

However, from the latter half of the 1950s in the socialist period (1924–89), cooperatives called *negdel*, which were modeled after *kolkhoz*, were introduced under the leadership of the Soviet Union. Each nomadic household that joined a *negdel* was ordered to supply their own cattle for *negdel* group ownership. The cattle were then grouped together, divided according to type of animal, and redistributed to each household for breeding. Due to this introduction of grouping and division of cattle breeding, the nomadic households were requested to breed only one kind of domestic animal. Although seasonal migration continued, the route was decided by the executives of the *negdel* and the range of movement became smaller. This system was counter to the tradition of the Mongolian nomads.

Disaster Resilient Cities
http://dx.doi.org/10.1016/B978-0-12-809862-2.00007-3

Photo 2.5.1 Nomads' life with animals.

The excessive control and extreme divisional breeding during the socialist period lacked proper use of grazing land, and resulted in a lot of damage by natural disasters. The evaluation that these bad effects were caused due to neglecting the traditional wisdom of nomads has been more or less established. The *negdels* collapsed after the 1990 democratization, and the nomads again started to keep five kinds of domestic animals, and regained the lifestyle of seasonal migration (Photo 2.5.1).

The excessive population concentration in Ulaanbaatar has brought various urban problems such as air pollution and traffic congestion. The redevelopment of Ulaanbaatar is necessary to solve these problems. Development of mines and inflow of foreign capital started to bring about big changes in the industrial structure and in lifestyle. In this situation, the question is how to utilize the "wisdom of resilience" that the nomadic society has fostered for future urban development and land planning.

2.5.2 *GER* DISTRICTS AND THE REDEVELOPMENT OF ULAANBAATAR

One of the key components of the wisdom of resilience in Mongolia is the movable-type house called a *ger*.

A *ger* consists of two center props, cornice-type wooden frames supporting the sides, and the felt that covers the whole structure. It takes only about 1 hour to construct. The entire weight is about 250 kg, and it is easy to carry a *ger* on camels' back or on a truck (Photo 2.5.2). Since it is covered with felt, it is quite warm even in winter if a stove is used in it. In summer, it is cool if the bottom part of the felt is opened for ventilation. A *ger* is used not

Photo 2.5.2 Caravan of nomads carrying a *ger* and housewares on the camels.

only for nomadic life; in the capital city of Ulaanbaatar, many people have lived in *gers* since olden times and have formed *ger* districts.

2.5.2.1 Capital Construction and *Ger* Districts

The capital of Mongolia was formed from the 17th century around a *ger* monastery, where the living Buddha of Mongolian/Tibetan Buddhism stayed. In those days, the capital was repeatedly moved until it was settled in its current position in 1855 [1]. Fig. 2.5.1 is a pictorial map of the capital around the beginning of the 20th century, depicting the palace for the living Buddha on the right (east side), while Gandan Monastery, built in the 19th century, can be seen on the left (west side). There are *ger* villages for priests surrounding Gandan Monastery. Those villages remain almost the same as they were in those days.

Fig. 2.5.1 Illustration showing Ulaanbaatar City in the early 20th century. *(Data from Atlas of Ulaanbaatar. Ulaanbaatar: State Geodesy and Cartography Office; 1990).*

The capital was renamed Ulaanbaatar in 1924 through the influence of the Soviet Union, and developed later as a modern city. In those days, workers were required for the city construction and the development of mining and industry, so the government invited nomads to Ulaanbaatar from the countryside. As a result, the city population increased rapidly [2].

On the other hand, the control system of socialism by *negdel* was reinforced, which made many nomads dissatisfied about being ordered to provide their cattle. At the same time, surplus nomad labor was produced. These people gathered in the city, causing a rapid population increase of Ulaanbaatar. Many apartment houses were constructed to support the rapidly increasing population, but there was always a shortage of apartments to meet the rapidly increasing population, and many citizens lived in *gers* (Photo 2.5.3).

When democratization and a market economy started in 1990, many laborers lost their jobs due to the factory closures and bankruptcies of companies during the socialist period. The functions of education and welfare, which were guaranteed during the period, were weakened. Vulnerability to disasters increased. When the nomads lost their all cattle after being hit by a *dzud*, they had no alternative but to go to the cities to find jobs. On the other hand, unemployed laborers in the city had to move to the countryside. This resulted in increased mobility of people.

Another factor that promoted the inflow to cities was the land privatization act, which was enacted in 2003. This act enabled Mongolian people to obtain a certain area of land free of charge in their permanent district. Since this privatization project was enforced in *ger* districts in Ulaanbaatar (and also

Photo 2.5.3 Apartment houses built during the socialist era and a *ger* district.

Photo 2.5.4 *Ger* districts extend behind buildings in 2014.

in *ger* districts in the provinces, or *aimag*, and their subdivisions, or *sum*), many people wished to obtain their land in the capital rather than in the countryside, and immigrants from the countryside increased. So, *ger* districts in Ulaanbaatar became more overcrowded, and they expanded in a disorderly way to the dangerous slopes and low marshes along the rivers (Photo 2.5.4).

When this kind of rapid population movement occurs, the existence of *ger* districts plays an important role as a first step to living in the capital. As long as they have a *ger*, they can somehow make a living. In *ger* districts, they can keep cattle in a *hashaa*, which is cattle enclosure fenced with boards, and partially maintain their nomad life.

However, the *ger* districts are not equipped with such infrastructure as hot-water piping for heating, and a water and sewage system. So, people have to burn coal to warm themselves in the extremely cold winter, which causes very severe air pollution that is said to be worse than in Beijing. Furthermore, since no sewage system is available, sanitary conditions have worsened and the soil has been contaminated.

Under these circumstances, an urban redevelopment plan started in 2013 to build apartment houses by removing *gers*. While this plan is unavoidable and necessary to solve the above-mentioned problems, it brings up the very difficult issue of how to maintain *ger* culture that is the basis of Mongolian way of life and thinking. There is an opinion among citizens that *gers* are not necessary in cities. However, since half of the population lives in Ulaanbaatar, the effect of their becoming unrelated to *gers* cannot be disregarded from the viewpoint of maintaining traditional culture.

Photo 2.5.5 *Ger* district in Ulaanbaatar City.

2.5.2.2 The Structure and Characteristics of *Ger* Districts

In the *ger* districts, residents' sites are surrounded by *hashaa* for each site of about 400–700 m². The arrangement of *hashaas* looks like cobwebs as they are placed next to each other (Photo 2.5.5). The houses in each *hashaa* are mostly *gers*, but they are being replaced by wooden or brick houses. Electricity is now supplied in *ger* districts, but water and sewage systems were still not available in 2014. No central heating system is available in the *ger* districts, and the *ger* residents use coal for heating and buy water from water-supply stations.

The *ger* residents' income is on the whole low, but there are some middle-class people. Among them are judges, scholars, and rich shop-owners. Some families are engaged in cattle breeding in *ger* districts in the suburbs. They are mostly immigrants from the countryside, or their second and third generations. There are not a few cases in which multiple families live together, building multiple *gers* and wooden houses in the same site. This case of sharing the same site is fairly popular, and the tradition of mutual assistance by nomads is maintained.

There are some people who live in *gers* in the *ger* districts in summer and move to apartment houses in winter. As soon as they find an advantageous place, they move. Depending upon the situation, they have the flexibility to change professions, houses, and cohabitants, and live on the site if it is not occupied, and build *gers* temporarily, even if it is somebody else's site. The generous atmosphere which allows this to happen is possibly due to the site sharing that is found in normal nomadic life.

These characteristics of movability, flexibility, sharing, and mutual assistance may be called "traditional wisdom" to maintain the resilience originating from nomadic life. Because the *ger* districts exist, traditional wisdom is also maintained even within cities [3].

2.5.3 REDEVELOPMENT PLAN FOR *GER* DISTRICTS AND THE ATTITUDES OF RESIDENTS

As mentioned earlier, *ger* districts are deeply rooted in the Mongolian tradition in particular, and the redevelopment plan does not proceed so easily. Ulaanbaatar City is planning to introduce technologies and knowledge for urban redevelopment from abroad such as from Japan or other Western countries, but many adjustments unique to Mongolia will probably be required. The following is the progress up to 2014.

2.5.3.1 Outline of the Redevelopment of *Ger* Districts

The redevelopment plan of the *ger* districts is a plan to combine the private land in *ger* districts by the unit of from several to dozens of sites, and build a condominium on the combined site. People who offer land are entitled to a room in a condominium. There is an alternative to move to another place without moving into the condominium after being paid for the land.

Ulaanbaatar City informed the citizens of the details of the redevelopment project of *ger* districts in their city brochure. According to the brochure, the redevelopment project will continue for the next 6 years, and infrastructure such as water, sewage, and a heating system is to be prepared, and the necessary laws and ordinances for the support and maintenance of the redevelopment are going to be introduced. The brochure urges the residents to participate positively in the redevelopment, and emphasizes that their opinions will be reflected and respected.

2.5.3.2 Opinions of the Residents Toward the Redevelopment Project of *Ger* Districts

In cooperation with the National University of Mongolia, in Nov. 2013, we conducted a questionnaire survey about the opinions of the residents living in the five *ger* districts: (1) the historically old *ger* district surrounding the Gandan Monastery, which is the base for Mongolian/Tibetan Buddhism (hereafter referred to as the Gandan *ger* district); (2) the *ger* district in the center of Ulaanbaatar; (3) the *ger* district in the suburbs of Ulaanbaatar; (4) a *ger* district with flood risk; and (5) a *ger* district that was decided to be rebuilt to

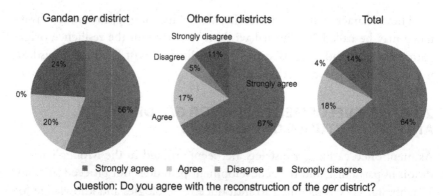

Question: Do you agree with the reconstruction of the *ger* district?

Fig. 2.5.2 Results of questionnaire in five districts of Ulaanbaatar in 2013.

apartments at the earliest stage and is the most advanced in the project. In the questionnaire, we asked residents of 100 families to inform us of their family composition, annual income, the comforts and complaints of living in a *ger* district, and how they felt about the district being rebuilt with apartment blocks.

First, let us see the results of the five *ger* districts. To the question, "How do you feel about *ger* districts being rebuilt with apartment blocks?," the answers were "Strongly agree: 64%," "Agree: 18%," "Disagree: 4%," and "Strongly disagree 14%" (Fig. 2.5.2). From these results, we see that many residents of the *ger* districts are for rebuilding with apartment blocks. The main reason for the agreement was "Rebuilding with apartment blocks equipped with a heating system will stop burning coal for heat, and solve the air pollution problem."

However, the Gandan Monastery *ger* district showed a different tendency from those of the other four districts, and the percentage was as shown in Fig. 2.5.2. According to this tendency, the percentage of the residents of the Gandan Monastery *ger* district who were for rebuilding was slightly smaller, and the percentage of the residents who are strongly against was more than twice that of the other *ger* districts. Gandan Monastery *ger* district showed a different development from the other districts in the next year's survey of the questionnaire.

2.5.3.3 Progress and Change in the Summer of 2014 (After 1 Year)

Rebuilding with apartment blocks in the Gandan Monastery *ger* district was canceled in 2014. It was decided to install water, sewage, and a hot-water heating system, leaving the *ger* district as it was, and the installation work has already started, and is scheduled to be completed by 2016.

In the previous year's questionnaire survey for the Gandan Monastery district, the percentage of people who were against rebuilding to apartments was higher than those in other districts. Those people who were against played a key role in the opposition movement. They selected a leader for the opposition group, began a campaign as a group, and submitted a suggestion to the city office.

The Gandan Monastery district has a long history since the establishment of Ulaanbaatar, and continues the original form of a Mongolian city. The residents well understand the significance of maintaining the *ger* district as it is as a historical preservation area. Furthermore, they are well aware that it is more profitable not to sell their land since the land price has soared due to its location in the city center.

In some other *ger* districts, the residents have already left, and the construction of apartments has already started (Photo 2.5.6). In other districts where it was decided to build apartments, the plan was stalled due to unsuccessful negotiations between the residents and the construction company. According to the residents, a district meeting was held to discuss rebuilding as apartments, but no opportunity was given to them to give their opinion on the specific plans. The rebuilding plan was to be given the top priority, reflecting the residents' opinions, but the plan was behind schedule as of Sep. 2014.

If the rebuilding plan proceeds further in the future, the *ger* districts, which have preserved the flexible life strategies of nomadic people, may disappear from Ulaanbaatar. However, it is also true that chaotic expansion of *ger* districts, and overpopulation have caused many serious urban problems, and the trend of rebuilding to apartments cannot be stopped.

Photo 2.5.6 Construction of apartment houses in the redeveloped *ger* district in 2014.

Is it not possible to redevelop the *ger* districts in such a way that they can hand down nomadic culture while still solving the various problems such as environmental damage, and yet obtain comfort of living? For example, how about utilizing the space that becomes available by rebuilding with apartment blocks so that *ger* districts can be well maintained for public use? Another idea is to make evacuation space for use in a disaster by arranging public *ger* space in succession in the urban area. *Ger* districts may also serve as tourism resources as well as places where children raised in the urban area can come into contact with the *gers*. In this way, the *ger* districts can contribute to maintaining the resilience particularly connected with Mongolia. What is important is to make these kinds of plans and implement them as planned while rebuilding *ger* districts as apartments.

2.5.4 TOWARD THE RESILIENCE STUDY IN MONGOLIA

Mongolia has been achieving rapid economic growth, and the situation of the society has been changing greatly. In this situation, the Mongolian people have come to realize that maintaining resilience is very important. They have faced natural disasters such as *dzud*, desertification and earthquakes, as well as urbanization issues. It is required that they should study the social structure and the land structure in future from the viewpoint of resilience.

First of all, measures to secure resilience can be found in various best practices, but, on the other hand, we can also learn from bad examples. It is necessary that we should introduce to Mongolia antidisaster measures already established in other countries, but we need to understand that those measures may have to be adapted to be compatible with the traditional culture. They may conflict with ideas for pursuing economic efficiency. So, we need to deepen our understanding by bringing together the knowledge of various fields, and by learning from traditional wisdom. In addition to an interdisciplinary approach, transdisciplinary projects in collaboration with the society as well as by the experts will be needed.

To promote international cooperation between Mongolia and Japan, the Cooperative Center for Resilience Research with Nagoya University will be established in the National University of Mongolia. The purpose is to study the Mongolian environment, disasters, and society from the viewpoint of attaching great importance to resilience and sustainability. The center is also intended to create a system that enables Mongolian youth to act more independently. Resilience is a very important concept to utilize the culture and tradition that are characteristic of an ethnicity, and to build a sustainable society. While developing the human resources to support resilience study, we must be engaged in a long-term project.

REFERENCES

[1] Baabar B. History of Mongolia. [Suhjargalmaa D, Burenbayar S, Hulan H, Yuya N, Trans.]. Kaplonski C, editor. University of Cambridge Cambridge: The White Horse Press; 1999.

[2] Ricardo N, Goldstein S. Urbanization and population redistribution in Mongolia. Honolulu, HI: East-West Center; 1994 p. 14–36.

[3] Ishii S, Suzuki Y, Inamura T. Grassland and the city. Fubaisha, Nagoya; 2015 [in Japanese].

CHAPTER 2.6

Decreasing Disaster-Recovery Capacity of Traditional Wooden Houses in Japan

Y. Kawazoe, M. Yoshitake
Institute of Industrial Science, The University of Tokyo, Tokyo, Japan

2.6.1 THE CHANGE IN PRODUCTION OF JAPANESE WOODEN HOUSES

The Japanese have lived in a culture that uses wood for many things, from houses and everyday items to fuel. The wooden buildings lining Japanese streets have been restored every time they were hit by natural disasters that could not be escaped geographically, or caught in a large-scale fire. Edo, which was called "Fire City" at that time, was hit by as many as 40 great fires. People engaged in the restoration work were carpenters and scaffold builders who had gained craftsmanship by division of labor. According to the traditional craftsman employment system that lasted until the Showa period, the master (builder) paid the same wages to his craftsmen even when they were not engaged in any particular work in an ordinary time, as long as they had been preparing building materials in their workshop. But in an emergency like a fire, those craftsmen dashed to the master to help him. As to the materials, a rich merchant used to store enough wood for a house at his lumber dealer, preparing for a possible fire. Here we shall see their way of thinking to secure redundancies on the assumption that there would be a great demand for work following a disaster. In order to recover ordinary life quickly after a disaster that would surely hit them, they had learned to keep enough technicians and materials on hand to reconstruct houses. As seen in a storehouse with thick plaster walls, they had also taken fire-proofing measures, using multiple layers of plaster on the wooden structure. On one hand, people had secured safety by preparing structure-based measures, and on the other hand, they kept enough stock in advance of possible damage by external forces. Was this not a resilient society that can be used as a reference in the present?

In the postwar period, the housing situation in Japan changed greatly. In a panic with a shortage of 4.2 million houses after the Second World

Disaster Resilient Cities
http://dx.doi.org/10.1016/B978-0-12-809862-2.00008-5

War, people worked hard without complaining to obtain their own house, supported by the "having our own houses" policy. In the 30 years after the war, the number of households and the number of houses were kept in equilibrium, and the necessary housing quantity was secured. However, the poor-quality houses that could barely be built with private funds were described as "rabbit houses" in the beginning. So, later, a change of housing policy from quantity to quality was advanced. In other words, it was promoted that poor-quality houses were to be rebuilt with a design of higher functions. Together with rapid economic growth, people's desire for good-quality houses enabled the housing industry to become great, causing the bubble economy [1]. We wonder why the planning aiming at more essential affluence over the long term could not have been made in the 1970s when the change from quantity to quality was sought. This is a reflection on the growth period, which must be succeeded by us living in the present age.

What was brought about in communities by the change in housing production in the postwar period? As one example, let us see Fig. 2.6.1 showing the results of the survey conducted in Katori City, Chiba Prefecture, a Tokyo suburb that has historical streets [2]. Of the houses built in 1967, by weighted percentage, 1% were made with imported materials, and 99% were made with domestic materials (45% were made with materials produced within a 10 km radius). Meanwhile, of the houses built in 2010, imported materials were used even for the main structural materials. In total, by weight percentage, imported materials constituted 21%. Materials were supplied by industrial sectors as follows. In 1967, by weight percentage, 56% of the materials came from the primary sector (forestry and agriculture) in the neighborhood, while in 2010, 94% of the materials came from the secondary sector (industry). Once, there was an economic cycle connecting the construction industry and the primary sector. At present, to shorten working hours at the construction site, industrial products are used wherever possible. Here, we see that postwar Japan's course with too much emphasis on the manufacturing industry is reflected in the change of houses. Sahara was once a town where a merchant culture that could have been superior to that of Edo prospered, but now the craftsmen who construct traditional houses do not have successors. Therefore, they had to invite specialized craftsmen from other regions to repair the houses considered to be important cultural heritage that were damaged by the Great East Japan Earthquake.

As materials used for housing are being replaced by industrial products that are not related to the culture of a community, craftsmen related with

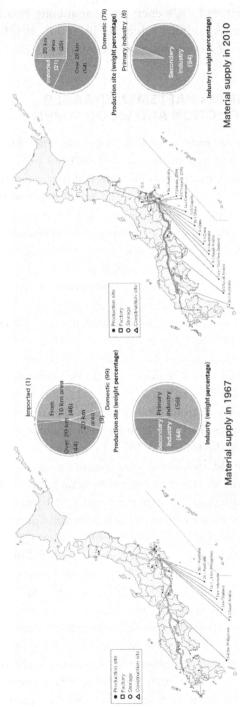

Fig. 2.6.1 Changing housing material supply, comparison of 1967 and 2010, Katori, Chiba Prefecture.

the culture disappear, and the scenery of a community becomes homogeneous all over Japan. As cultural values are being lost, it is feared that the automatic resilience of a community has decreased.

2.6.2 DECREASE IN CRAFTSMEN ENGAGED IN HOUSE-PRODUCTION AND WOOD SUPPLY

Decreased housing recovery capability of communities that was brought by the change of house-production can be seen from yearly changes in the balance between the demand of workers needed for house-production and supply of wood. According to the Housing Starts Statistics, the number of newly started houses has been decreasing since 1970, recovered a little in 1990, but again started to decrease, reaching 460,000 houses in 2010. In a shrinking society, the demand for houses is expected to keep on decreasing.

Fig. 2.6.2A shows the balance of supply and demand of the number of carpenters against the number of new housing construction starts [3,4]. The carpenter sufficiency rate decreased nationally from 1960 to 1970 because of the carpenter shortage caused by increased demand for housing. From 1980 onwards, the sufficiency rate was in equilibrium at 2.0 because the decrease of housing demand and the decrease of carpenters were happening simultaneously. The change of sufficiency rate by prefecture shows that in the three metropolitan areas of Tokyo, Aichi, and Osaka, the number of carpenters started to decrease from 1960, which was a peak year, and there was a shortage of carpenters in 1970. However, the number of carpenters increased up to 1980 in the Tohoku and Kyushu regions. In these 20 years, the ever-increasing demand for labor in urban regions is supposed to have been supported by migrant workers from the peripheral regions. Due to rapidly increasing demand for housing after the Great East Japan Earthquake, the sufficiency rate has started to decrease notably in some prefectures. At present, labor costs have soared greatly because of the shortage of construction technicians who are engaged in housing production, and unsuccessful bidding for public works has continued. The structural issue of a shortage of technicians that we have in everyday life has surfaced due to the suddenly increased demand for housing after the disaster.

As to the supply of wood, expanded afforestation which could be described as a reaction to the huge wood demand in the postwar period, the importing of a great amount of low-priced wood, and the spread of new construction materials have distorted the balance between supply and demand of wood in the domestic market. The self-sufficiency rate of wood,

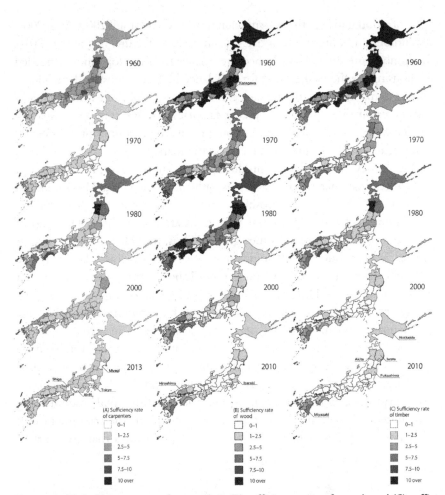

Fig. 2.6.2 (A) Sufficiency rate of carpenters, (B) sufficiency rate of wood, and (C) sufficiency rate of timber.

which was once nearly 100%, hit bottom at 18.2% in 2000. At present, the rate has recovered a little to 27.9% because domestic conifers started to be used as plywood, and regionally produced wood was re-examined for use. The self-sufficiency rate of construction materials used in a region can be captured in two stages of lumber amount and log amount. The balance between demanded amount and supplied amount was calculated as the wood self-sufficiency rate [4,5].

Fig. 2.6.2B shows the lumber amount. In 1960, the only area where the sufficiency rate was below 1.0 was Kanagawa Prefecture, which is on the outskirts of Tokyo. Tokyo produced a great amount of lumber as an accumulation place of wood, and kept the self-sufficiency rate over 1.0 until 2000. After 2000, the lumber amount started to decrease in each prefecture. In the year 2010, the lumber amount nationally became smaller than the log amount needed for housing. As the background for imbalance between housing demand and lumber demand, we see the influence of large-scale laminated wood factories, and increased direct supply to precut factories through trading companies. The number of newly built houses from precut factories amounts to 538,000 in 2013, and has had a great influence on the market trend.

Fig. 2.6.2C shows the log production amount. In 2010, the number of prefectures that met the demand was only nine, including Hokkaido and Miyazaki. In Ibaraki and Hiroshima prefectures where they have trading ports, imported wood and production items are accumulated. The ratio of imported materials used becomes greater. Thus, the self-sufficiency rate of lumber is high, but the self-sufficiency rate of logs is low. Nationally, the self-sufficiency rate is relatively high in the Tohoku and Kyushu regions, but in the central part of Honshu where there is a lot of economic accumulation, no prefecture can produce enough logs to match the required wood demand.

Materials and manpower that are not available in an area have to be obtained from outside the area (either from outside the prefecture or from abroad). The money spent to cover the shortage will flow out to areas producing a surplus. We defined this cost for covering the shortage, or profit from surplus, as "housing resilience," and calculated the total amount using a weighted calculation of materials and manpower as the investment cost (unit cost). Fig. 2.6.3 shows that among the prefectures that are short of

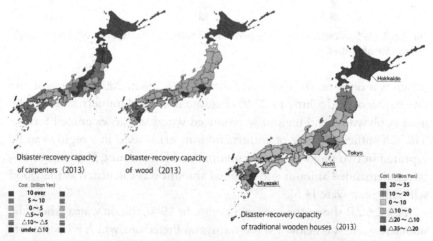

Fig. 2.6.3 Disaster-recovery capacity of traditional wooden houses in 2013.

housing resilience, the shortage in 2013 was 30 billion for Tokyo and 29.1 billion yen for Aichi. Even in the material cost only, four prefectures, including Tokyo, have shortages amounting to over 10 billion yen. Meanwhile, the prefectures that are highly resilient are Hokkaido (30 billion yen) and Miyazaki (24 billion yen). Prefectures with less resilience must always be aware that they have to depend on other regions both in ordinary times and in time of disaster. In order to restore the production capacity of wooden houses, the attitude to grow both human resources and materials following a long-range plan is indispensable.

2.6.3 PRODUCTION CAPACITY OF WOODEN HOUSES NEEDED AT THE TIME OF A DISASTER

In Yamakoshi village, which was severely damaged in the Niigata Prefecture Chuetsu Earthquake in 2004, there was normally one carpenter who was engaged in new construction and repairs in almost every one of the settlements. The existence of these carpenters was indispensable in the Housing Reconstruction Project. According to the project, the administrative offices, construction companies, architects, etc. worked together, and eventually by the techniques of local carpenters, houses were reconstructed with the traditional wooden framework method that can be maintained and managed. The reconstructed houses in the mountains are merged into the scenery quite well (Photo 2.6.1) [6]. If those houses had been built with

Photo 2.6.1 Restoration of public housing in Kyu-Yamakoshi-mura, Niigata Prefecture (Nov. 20, 2006, by Ito K.).

the prefabricated method using steel frames, the time needed for the reconstruction process might have been much shorter. However, if that had been
the case for Yamakoshi village with its scenic beauty, the use of modern materials would have fatally destroyed the scenery. Through the construction
of houses, carpenter jobs in the villages can continue, their successors can be
trained, and the scenery of the village will be maintained. The author thinks
that this kind of small cycle is essential for the maintenance and reconstruction of the village units in rural regions and fishing villages.

Sumita Town in Iwate Prefecture provided people with 110 temporary
wooden houses for their own project after the Great East Japan Earthquake
in 2011 (Photo 2.6.2). This project was also made possible because the ter-
tiary sector started to sell houses in 1982 with the scheme to grow trees and
use them with all town people united [7]. Temporary wooden houses using regional materials were also provided from other regions of Tohoku. The
Tohoku region still has not lost production capacity of wood, and the maintenance demand for houses is high because of heavy snowfall. The decrease of
people engaged in production is more moderate than that in other districts.
This is also a good example of where production capacity of the area was fully
made use of in responding to a disaster.

While the demand for wood in an ordinary time decreases, how should
we maintain production capacity of houses in a community? People in
Kaneyama Town in Yamagata Prefecture have been engaged for more than
30 years in maintaining wood production and housing supply as the basis

Photo 2.6.2 Temporary housing using wood materials from Sumita-cho, Iwate
Prefecture (Oct. 18, 2011, by Sugawara D.).

of community building. Kaneyama Town is a small agricultural village surrounded by mountains with a population of less than 6000, and is famous for producing cedars with a standard to cut trees of more than 80 years old. In the 1970s, the town already had various issues such as under-population and need for community revitalization. Therefore, in the townscape ordinance issued in 1986, the "formation standard" to promote building the "Kaneyama house," which is characteristic of the townscape, was established, and houses that matched the standard were subsidized with the maximum amount of 800,000 yen. At the end of 2014, the number of subsidized houses reached 1584, the total subsidy amount exceeded 230 million yen, and the total project amount reached 9.3 billion yen. This subsidy system was established with the common understanding that the townscape is not owned by each individual, but is a common asset of the townspeople. This system has also led to securing orders for local constructors, and maintaining their quality of work. (Local constructors received 88% of the orders for new houses, and 93% of the orders for enlarged and remodeled houses.) This system enables the community craftsmen to maintain their work and to bring up their next generation [8] (Photo 2.6.3). Continuing to keep the beautiful townscape and the supporting industry by ordinance will surely become a big help to the community in case of a possible disaster. It will eventually lead to the enhancement of resilience of Japan as a whole for a community to become independent with civic pride by a long-range project, as seen in Kaneyama Town.

Photo 2.6.3 Typical Kaneyama house in Kaneyama-cho, Yamagata Prefecture (Aug. 27, 2015).

REFERENCES

[1] Honma Y. Jutaku. Sangyo no Showa shakaishi. [Housing—social history of industry in Showa series no. 5]vol. 5. Tokyo. Nihon Keizai Hyoronsha; 2001. 307 pp.

[2] Yoshitake M, Kawazoe Y. "Building Mile" of small-scale house construction system—study in Sawara-area, Katori, Chiba. J Archit Plann Archit Inst Jpn 2015;80(707):1–8 [in Japanese].

[3] Statistics Bureau. Population census. 1960–13.

[4] The Ministry of Land, Infrastructure, Transport and Tourism. Survey on construction works. 1960–13.

[5] Ministry of Agriculture, Forestry and Fisheries. Statistical yearbook of Ministry of Agriculture, Forestry and Fisheries. 1960–13.

[6] Takeda K. Fukkojutaku ni tsuite kyuyamakoshi mura no jutaku saiken shien [Restoration public housing in Kyu-Yamakoshi village]. Jutaku, Jpn Hous Assoc 2012;61:52–7 [in Japanese].

[7] Mizuno Y. Chikizai katsuyo gata no mokuzou kasetu jutaku [Wooden temporary houses by use of local material]. Wood Ind Wood Technol Assoc Jpn 2011;66:535–6 [in Japanese].

[8] Shinohara O, et al., editors. Kono machi ni ikiru—seiko suru machizukuri to chiki sai-seiryoku. [Living in the town—successful town design and area regeneration] Tokyo. Shokokusha Publishing; 2013. p. 8–64 [in Japanese].

Learning Disaster Resilience from the Great East Japan Earthquake

KEY MESSAGES

- Chapter 3.1 examines damage maps produced at the time of a large-scale disaster and their roles. Materials for the examination are our tsunami inundation area map that the authors and others prepared using posttsunami aerial photographs, and mapping results of the tsunami run-up heights measured using digital elevation model afterward. Based on these examinations, this section reveals the remarkable spatial variation observed in the range and height inundated by the tsunami, and discusses its scientific significance and lessons for disaster management.
- For speedy recovery from a natural disaster, it is necessary to estimate in advance the amount of material stock that may lose its capacity to provide intended services as a result of the disaster, and the distribution of such stock. Understanding the amount of such stock to be generated will encourage proper disposal or reuse, as well as discussions on measures to improve regional resilience before and after the disaster.

- A long-term decline in the living standard of residents associated with a large-scale disaster may cause health impairment or even death. Measures to curb the decline in Quality of Life (QOL) after a disaster and expedite recovery are therefore important. This section proposes a methodology capable of evaluating chronological changes in QOL from occurrence of a disaster to restoration stages in units of small districts. Using this methodology, the areas hit by the Great East Japan Earthquake are evaluated to clarify quantitatively the differences in recovery processes among areas and the differences in the degree of deterioration/recovery, depending on differences in the initial level of infrastructure development.

CHAPTER 3.1

Understanding Spatial Variations of Tsunami Flooding

N. Sugito*, Y. Suzuki†, N. Matsuta‡
*Faculty of Humanity and Environment, Hosei University, Tokyo, Japan
†Disaster Mitigation Research Center, Nagoya University, Nagoya, Japan
‡Graduate School of Education, Okayama University, Okayama, Japan

3.1.1 DAMAGE MAPS AT THE TIME OF A LARGE-SCALE DISASTER

When a large–scale disaster occurs, it is difficult to comprehend its entire picture. What is going on, where it is happening, and what places are most seriously damaged cannot be known quickly. It therefore becomes important to describe the situation of the disaster with various scales of maps. In the immediate aftermath of the Great East Japan Earthquake in 2011, analysis using remote sensing imagery, such as aerial photographs and satellite images, as well as meticulous on-site surveys, was conducted, and various damage maps were developed.

Contents and accuracy/resolution required for a damage map vary greatly depending on when it is developed: in the rescue stage, restoration stage, or revival stage. In the rescue stage immediately after the disaster, comprehending the status of damage of the seriously affected areas where the traffic is blocked is important. In the restoration stage, maps that are useful for developing strategies to conduct restoration activities are needed. In the revival stage, more accurate information on locations as well as altitudes is also necessary for future planning.

This section examines what is demanded of damage maps and what roles they may play in ever-changing situations. It is also important to discuss how the government agencies and research institutions in charge of disaster management should share roles in developing the maps.

The role of a damage map is not only to contribute to rescue, restoration, or revival. A damage map plays a significant role in communicating the memory of a disaster to subsequent generations as a "disaster history." It may serve as the basis for developing hazard maps in the future. It also provides material for discussing how we should relate "prediction uncertainty," an issue presented by the Great Earthquake, to disaster management.

Disaster Resilient Cities
http://dx.doi.org/10.1016/B978-0-12-809862-2.00003-6

75

There are diverse forms of earthquakes and tsunamis, and the idea that a unique form of earthquake or tsunami only repeats is a mistake. In the case of the Great East Japan Earthquake, for example, some places were hit by a tsunami similar to the Meiji and Showa Sanriku tsunamis. At other places, on the other hand, the level of the tsunami was lower or higher than in the past. At some other places, the tsunami was not very high but its resultant inundation area was extraordinarily wide. There are also valleys next to each other that experienced tsunamis differing greatly in height. Detailed examination of why each of these phenomena occurred helps us understand how the tsunami run-up may vary according to the characteristics of the source fault of the earthquake. If the variation in predictions and its cause are identified, we may positively redefine the "prediction uncertainty" as meaningful uncertainty, instead of seeing it just as being vague and unreliable. Accepting disaster predictions including uncertainty will lead to avoidance of unanticipated damage, resulting in improved resilience.

This is how the disaster history map of the recent great earthquake should be used, and relevant attempts have just been launched. The following are examples of such attempts, as well as the roles of damage maps.

3.1.2 TSUNAMI MAPPING BY THE ASSOCIATION FOR JAPANESE GEOGRAPHERS IMMEDIATELY AFTER THE EARTHQUAKE

Immediately after the Great East Japan Earthquake took place, the media announced that there had been tremendous damage caused by the tsunami in broad areas from Tohoku to Kanto, though it took a long time to obtain detailed information on locations and damage of the affected areas. It is usually the case that the more serious damage is, the more difficult it is to establish the status of the damage. However, we soon found that the Geospatial Information Authority of Japan (GSI) had started taking aerial photos immediately after the earthquake and began to release them on Mar. 13. We geographers, who are used to stereo-pair interpretation of aerial photos in active fault studies, etc., responded almost like a conditioned reflex to them and quickly began to observe them.

On Mar. 18, the GSI developed a tsunami inundation map on a scale of 1:100,000 and revealed a rough figure of the inundated area. The limitation on accuracy, however, made it uncertain when a more detailed inundation map would be issued by the GSI. As the aerial photos showed many destroyed and isolated areas, it seemed to be urgently necessary to produce a detailed map for rescue operations. We therefore decided to produce a tsunami damage map on a scale of 1:25,000 while keeping in contact with the GSI [1].

On Mar. 24, 11 members from universities all over Japan gathered at Nagoya University, where the results of aerial-photo stereo-pair interpretations brought in by each member were examined and cross-checked, and the resultant findings were summarized. On Mar. 28, we, the Tsunami Damage Mapping Team of the Association of Japanese Geographers, released on the Web "Maps of the Area Hit by the Tsunami of 11 March 2011, Northeast Japan" [2] (http://danso.env.nagoya-u.ac.jp/20110311). We initially intended to draw maps of only the tsunami inundation area, though we subsequently decided to separately show the heavily damaged areas within the inundation area (Fig. 3.1.1).

We initially provided scanned handwritten maps and later converted them into GIS data, which was then released on Apr. 8 on the GSI's Digital Japan Web System and the "maps by e-community platform"

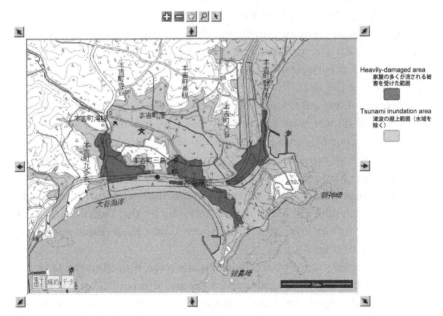

Fig. 3.1.1 Example of "Maps of the Area Hit by the Tsunami of 11 March, 2011, Northeast Japan" (Oya coast, Motoyoshi-cho, Kesennuma City), shown on the Digital Japan Web System. *(Data from Matsuta N, Sugito N, Goto H, Ishiguro S, Nakata T, Watanabe M, et al. Significance and background of mapping the area hit by the tsunami on March 11, 2011, northeast Japan. EJ GEO 2012;7:214–24 [in Japanese with English abstract]).*

of the National Research Institute for Earth Science and Disaster Prevention. By overlaying a map converted into GIS data on the digital elevation models (DEMs) of the Digital Geospatial Data, the tsunami run-up height can be roughly figured out. This approach led to the development of the Spatial Variation Map of Tsunami Run-up Heights for the 2011 off the Pacific Coast of Tohoku Earthquake, details of which are explained later.

In disclosing digital information, we were particularly careful not to allow the data to be used in unintended ways. Once a map is converted into digital data, it can be overlaid on any large-scale map, which is associated with the risk of it being used in discussions as material with accuracy higher than its actual locational accuracy. We therefore decided not to allow public access to original GIS data (files in the shape or KML format), but to show them only as images on the Digital Japan Web System or the e-community platform. Moreover, we restricted enlargement of the maps in such a way that relevant tsunami information will disappear if the data is enlarged to a scale larger than 1:25,000. Meanwhile, however, we decided to accept applications for use of GIS data from institutions and organizations having a clear purpose of use, such as for developing rescue strategies, analyzing damage, and creating materials for education for disaster mitigation, and to provide them with the data on condition that they were fully aware of the accuracy limits of locational information of the data.

After Apr. 8, we continued to revise the tsunami damage maps. As additional aerial photos were provided by the GSI and also photos of some areas were offered by some private companies for purchase, we expanded the target range of our analysis. For the area around the Fukushima Daiichi Nuclear Power Station, of which aerial photos could not be taken because of the nuclear power accident, we used satellite photos provided by Google Earth. Consequently, the Maps of the Area Hit by the Tsunami for areas from central Aomori Prefecture to northern Chiba Prefecture were completed on Sep. 9 (Fig. 3.1.2).

Upon receiving aerial photos with higher resolution from the GSI, we conducted detailed analysis again, and for the area from northern Fukushima Prefecture to the north, drew the tsunami run-up lines on ortho-air photos at a scale of 1:10,000 and converted it into GIS data. This work helped to improve substantially the recognition accuracy and the locational accuracy of the maps. On Dec. 11, the 2011 final edition of the Maps of the Area Hit by the Tsunami was released.

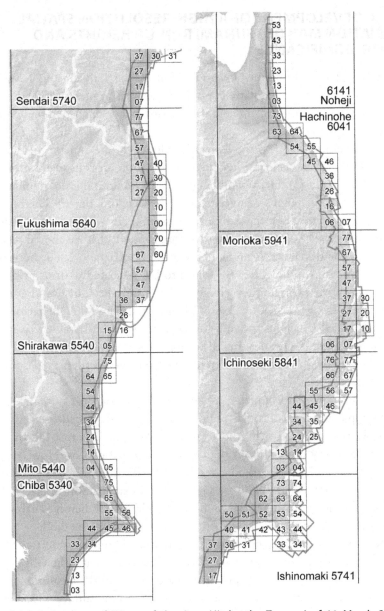

Fig. 3.1.2 Index map of "Maps of the Area Hit by the Tsunami of 11 March 2011, Northeast Japan." *(Data from Sugito N, Matsuta N, Goto H, Kumahara Y, Hori K, Hirouchi D, et al. Criteria for detection of tsunami-inundation area associated with the 2011 off the Pacific Coast of Tohoku Earthquake, Northeast Japan, based on air-photo stereo-pair interpretation. J Jpn Soc Nat Disaster Sci 2012;31:113–25 [in Japanese with English abstract]).* The bold-gray (1) polygons and (2) ellipses are indicative of the areas interpreted using (1) post-tsunami stereo-pair aerial photos taken by the GSI and (2) satellite images of Google Earth, respectively. The base maps are based on the GSI Data.

3.1.3 DEVELOPMENT OF A HIGH-RESOLUTION SPATIAL VARIATION MAP OF TSUNAMI RUN-UP HEIGHTS AND THEIR SIGNIFICANCE

After completing the Maps of the Area Hit by the Tsunami of 11 March 2011 described above, we started searching for ways to present information regarding the tsunami run-up heights on maps in a detailed, comprehensive manner, using our maps.

The heights related to the tsunami include: (1) the tsunami heights at coastal lines; (2) the tsunami heights in inundation areas (tsunami inundation heights); and (3) the tsunami heights at inland limit lines of inundation (tsunami run-up heights). Among them, the tsunami run-up heights are mostly referred to when speaking of a disaster, such as "The tsunami came up to a height of XX meters." We therefore thought that we should record accurately the tsunami run-up heights for future generations. Furthermore, the tsunami run-up heights were found to be associated with surprisingly broad spatial variation, as they greatly differ among bays, or even between valleys next to each other. In areas of plains, the tsunami flow behavior was also influenced by slight ups and downs of land or the layout of rivers. It therefore seemed quite meaningful to draw a spatial variation map of tsunami run-up heights as the first step to figure out such complicated tsunami flow behavior.

The attempt of recognizing a tsunami inundation area by employing stereoscopic viewing of aerial photos was almost unprecedented. It was therefore necessary to verify the accuracy of this method first [3]. In fact, there were many doubts about the capacity of aerial photos (How far can we see with aerial photos?) The results of verification, however, demonstrated their effectiveness. This method enabled examination of places that are inaccessible by on-site survey, and speedy and comprehensive analysis of broad areas, which are highly attractive advantages of the method.

Next, to find the tsunami heights at the inland limit lines on the Maps of the Area Hit by the Tsunami, we needed DEMs. Recently, the GSI began to provide 5 m mesh DEMs produced based on LiDAR measurement, a technique to measure heights by irradiating the ground surface with a laser from an airplane. We obtained and used the high-accuracy elevation data of GSI measured after the Great Earthquake [4].

In fact, among the data we urgently obtained immediately after the earthquake, the ortho-air photo used as the base map for analysis contained distortions, so we had to correct them. We also removed data for places whose geographical conditions made analysis difficult, such as steep cliffs covered with vegetation where it was impossible to identify accurately the tsunami heights at the inland limit line, and provided other adjustments. As a result of these efforts, in Mar. 2014, the "Spatial Variation Map of Tsunami Run-up Heights for the 2011 off the Pacific Coast of Tohoku Earthquake" was completed [5]. Maps for the northern and southern areas were also later developed [6].

Fig. 3.1.3 shows an example. The tsunami run-up heights are consecutively presented, so that we can see locational variation at a glance. Since our work, however, was not able to identify the tsunami heights at coastal lines and the tsunami heights in inundation areas, we put data of the Tohoku Tsunami Joint Survey (TTJS) Group [7] in the map.

The TTJS Group, consisting of over 300 members, shared the results of field measurement of the tsunami inundation heights or tsunami run-up heights and promptly released them on the Web, contributing to the subsequent establishment of the Tsunami Photo Archive and the Japan Tsunami Trace Database [8]. These data also played an important role in estimating the tsunami scale, as described in Chapter 4.1.

3.1.4 UTILIZATION OF TSUNAMI HAZARD MAPS IN THE FUTURE

The "Maps of the Area Hit by the Tsunami of 11 March 2011, Northeast Japan" and the "Spatial Variation Map of Tsunami Run-up Heights for the 2011 off the Pacific Coast of Tohoku Earthquake" presented above are both objective disaster history maps of the Great East Japan Earthquake. They are expected to be utilized as basic data in developing revival plans for the future or as teaching materials for education for disaster mitigation, in addition to the roles these maps played in rescue or restoration stages.

They may also serve as basic data in examining why tsunami run-up heights vary greatly among different places. In the case of the Great East Japan Earthquake, the tsunami anticipated in previously prepared hazard maps greatly differed from the actual tsunami that hit some places, which is pointed out to be a major cause of extensive damage. It is therefore necessary to re-examine the method of producing tsunami hazard maps in the future. The Spatial Variation Map of Tsunami Run-up Heights we have produced should be utilized as important data for discussing how accurately the tsunami run-up behavior can be simulated.

Meanwhile, we also conducted a survey on the seafloor landform around the Japan Trench. The results of the survey include a finding that there is

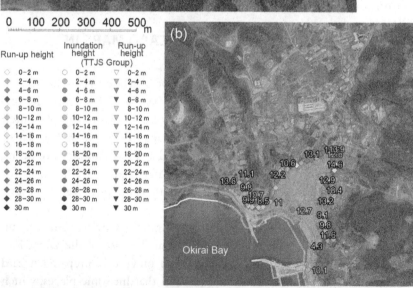

Fig. 3.1.3 Examples of "Spatial Variation Map of Tsunami Run-up Heights for the 2011 off the Pacific Coast of Tohoku Earthquake." *(Data from Sugito N, Matsuta N, Ishiguro S, Uchida C, Senda Y, Suzuki Y. Spatial Variation of Tsunami Run-up Heights for the 2011 off the Pacific Coast of Tohoku Earthquake, based on GIS analysis of the tsunami-inundation-area data and digital elevation model. J Geogr (Chigaku Zasshi) 2015;124:157–76 [in Japanese with English abstract]).* (a) Yoshihama, Sanriku-cho, Ofunato City; (b) Nakasakihama, Okirai, Sanriku-cho, Ofunato City. Inundation heights and run-up heights shown by circles and inverted triangles, respectively, are after TTJS Group [7]. The base maps are ortho-photos issued by the GSI.

an active submarine fault that was presumably displaced at the time of the Great East Japan Earthquake. Some explain this fault as a splay fault from the plate boundary. Many other active submarine faults that might be related to historical earthquakes were also discovered. Since each of these active submarine faults may become a source of tsunami waves, conducting tsunami simulations taking these faults into consideration in comparison with the historic tsunami run-up heights may help us understand the diversity of earthquakes occurring along the Japan Trench. Through such discussions, we will seek a framework of disaster prevention measures that pays attention to diversity in how an earthquake occurs.

REFERENCES

[1] Matsuta N, Sugito N, Goto H, Ishiguro S, Nakata T, Watanabe M, et al. Significance and background of mapping the area hit by the tsunami on March 11, 2011, Northeast Japan. EJ GEO 2012;7:214–24 [in Japanese with English abstract].

[2] Tsunami Damage Mapping Team, The Association of Japanese Geographers. Maps of the Area Hit by the Tsunami of 11 March 2011, Northeast Japan: 2011 final edition, http://danso.env.nagoya-u.ac.jp/20110311; [accessed 02.26.16]. 2011 [in Japanese with English abstract].

[3] Sugito N, Matsuta N, Goto H, Kumahara Y, Hori K, Hirouchi D, et al. Criteria for detection of tsunami-inundation area associated with the 2011 off the Pacific Coast of Tohoku Earthquake, northeast Japan, based on air-photo stereo-pair interpretation. J Jpn Soc Nat Disaster Sci 2012;31:113–25 [in Japanese with English abstract].

[4] Sugito N, Matsuta N, Ishiguro S, Uchida C, Senda Y, Suzuki Y. Spatial Variation of Tsunami Run-up Heights for the 2011 off the Pacific Coast of Tohoku Earthquake, based on GIS analysis of the tsunami-inundation-area data and digital elevation model. J Geogr (Chigaku Zasshi) 2015;124:157–76 [in Japanese with English abstract].

[5] Matsuta N, Sugito N, Ishiguro S, Sano S, Uchida C, Suzuki Y. Spatial Variation Map of Tsunami Run-up Heights for the 2011 off the Pacific Coast of Tohoku Earthquake: 1:25,000-scale compilation map. Disaster Mitigation Research Center, Nagoya University; 2014 [in Japanese].

[6] Sugito N, Matsuta N, Ishiguro S, Sano S, Uchida C, Suzuki Y. Spatial Variation Map of Tsunami Run-up Heights for the 2011 off the Pacific Coast of Tohoku Earthquake: 1:25,000-scale maps of central-southern Aomori Prefecture and central Fukushima to northern Chiba Prefectures. Disaster Mitigation Research Center, Nagoya University; 2015 [in Japanese].

[7] The 2011 Tohoku Earthquake Tsunami Joint Survey (TTJS) Group. http://www.coastal.jp/#jt/index.php; [accessed 09.19.13].

[8] Sato S. Characteristics of the 2011 Tohoku Tsunami and introduction of two level tsunamis for tsunami disaster mitigation. Proc Jpn Acad Ser B 2015;91:262–72.

CHAPTER 3.2

Understanding Tsunami Damage as "Lost Stock"

H. Tanikawa, K. Sugimoto
Graduate School of Environmental Studies, Nagoya University, Nagoya, Japan

3.2.1 IMPORTANCE OF UNDERSTANDING "LOST STOCK"

As some years have passed since the devastating Great East Japan Earthquake, reconstruction of the affected area is being discussed. Recovery or reconstruction of social activities is closely related to construction of buildings and infrastructure in the affected area. Therefore, a proper reconstruction plan that refers to the conditions of such buildings and infrastructure before the disaster, while also taking a long-term perspective, is required. In the course of restoring urban functions that support people's lives, an enormous amount of materials, including construction materials, will be needed. Production of construction materials, such as cement and steel, is associated with energy consumption and CO_2 emissions, and the future burden on municipalities should also be considered to maintain or renew the constructed facilities. Through past growth of cities, a large amount of construction materials has been accumulated in the form of buildings and roads, which have been utilized as a stock to provide social services. It is therefore necessary not only to understand the material amount of the constructions in the affected area but also to identify the amount of social services and social activities generated from such stock.

Social infrastructure or buildings generate services by forming a network. Constructing a road or a sewer alone, for example, is of no use. Similarly, at the time of a disaster, once a road is cut off, its related road network will lose all of its capacity to provide services. The same is true of buildings. If the upper structure of a building is swept away by a tsunami, the remaining lower base structure is unable to generate any service. Thus, quantitatively identifying the accumulated materials that will lose their functions as a result of a disaster, or "lost material stock," and having a clear image of it, seem useful in developing specific reconstruction plans to improve resilience after the disaster.

Disaster Resilient Cities
http://dx.doi.org/10.1016/B978-0-12-809862-2.00010-3

Here, the term "lost stock" is defined as the material weight of constructions that have lost the capacity to provide the services that they were supposed to provide, due to some damage. The benefits of estimating lost stock are: (1) to restore the amount of social activities held by the affected area before the disaster, the amount of constructions and infrastructure that originally supported such activities, or the baseline amounts of material necessary for reconstruction of the area, can be indicated; (2) spatial distribution of the quantity and quality of materials constituting the constructions and infrastructure before the disaster can be presented in maps; and (3) since classification by vertical location, such as above ground or underground, is applicable to this approach, the stock remaining after the disaster can be indicated, enabling determination of the amount of construction materials to be recovered in the case of building houses in the same place as before the disaster using their remaining base structures. Moreover, after clarifying the quality and quantity of construction materials to be accumulated as social infrastructure facilities or buildings (material stock) and the effects (services) of such facilities on society, it is important to maintain for the long term a highly efficient stock in terms of cost efficiency, production efficiency, environmental efficiency, and various other aspects, which will contribute to the improved long-term sustainability.

Section 3.2.2 describes the method of estimating the amount of materials accumulated as constructions. The amount of accumulated materials is calculated by multiplying the scale value measured by using the spatial information of each construction by the value of material intensity (the amount of construction materials applied per unit area). Section 3.2.3 shows—targeting the area affected by tsunamis in the Great East Japan Earthquake—how the amount of lost stock is estimated by combining the spatial information of the accumulated materials constructed and the tsunami inundation data. Finally in Section 3.2.4, Map Layered Japan, a website to deliver the results of the lost stock estimation, is described. Map Layered Japan can display its data overlaid with other geographical information and is therefore expected to be useful not only to government organs and researchers, but also for environmental and disaster management activities at the individual level.

3.2.2 A METHOD FOR ESTIMATING THE AMOUNT OF MATERIAL STOCK

To estimate the amount of lost stock, we first need to estimate the amount of materials currently accumulated in the form of constructions (material stock) in society. Therefore, the amount of material stock of each construction is calculated as:

$$\text{material stock} = \text{scale of construction} \times \text{material intensity},$$

where the scale of construction refers to the total floor area if it is a building, or the area obtained by multiplying the width by the length of the road if it is a road. In Tables 3.2.1 and 3.2.2, the values of material intensity, organized by our laboratory according to the type of constructions, the type of materials, and the time of construction, are used [1,2]. Material intensity can also be classified according to the upper structure or base of a building, which enables estimation of the amount of material stock separately for the building's upper structure and the base. As a result, the amount of construction materials avoided when building a house, etc. at the same location before the disaster using the remaining foundation, or the amount of material stock for the upper part of a building that was swept away by a tsunami, can be estimated.

For estimation of the amount of material stock for buildings, the building data of Zmap TOWN II provided by ZENRIN Co., Ltd. is used. In Zmap TOWN, the form of each building is recorded as spatial information with its attribute information—such as the purpose of use, number of floors, and name of the building—attached. The building area of each building is obtained based on the form of the building, and the total floor area can be calculated by multiplying the building area by the number of floors. Meanwhile, since Zmap TOWN lacks sufficient information regarding the number of floors, structure, and total floor area of buildings, we set these values by using the Housing and Land Survey results or employing a correction formula. For more details, refer to Tanikawa et al. [3]. For roads, on the other hand, the road (route) data of national geographic digital information contained in the ArcGIS Data Collection Standard Pack provided by Esri Japan Corporation is used. Spatial information of road networks is stored along with their attribute information such as the road type, width, and name.

Fig. 3.2.1 shows the spatial distribution with regard to the material stock of buildings and roads in the Tohoku region, represented in units of the Basic Grid Square (approx. 1 km). The material stock exists in 36,105 of the grid squares in the figure and about half of them, or approx. 20,000 grid squares, have a relatively small amount of stock of 10,000 ton/km^2. This indicates that grid squares with an extremely large amount of stock are located mostly in urban centers of big cities, such as Sendai, Tsukuba, and Fukushima.

Table 3.2.1 Material intensities of buildings classified by structure (unit: kg/m²)

Building type	Layer type	Materials									
		Aggregate	Cement concrete	Mortar	Wood	Glass	Ceramic	Steel	Aluminum	Other	Total
Wooden (W) house	Building	–	–	3	88	5	52	2	2	32	184
	Foundation	78	221	–	–	–	–	5	–	–	304
Steel-based, 1-story (S1) building	Building	–	–	67	8	0	2	132	0	25	234
	Foundation	339	584	–	–	–	–	7	–	–	930
Steel-based, 2-story (S2) building	Building	–	–	109	20	3	1	104	2	22	261
	Foundation	100	587	–	–	–	–	14	–	–	701
Steel-based, 3-story (S3) building	Building	–	–	143	4	1	1	165	1	–	315
	Foundation	214	416	–	–	–	–	13	–	–	643
Reinforced concrete (RC) building	Building	–	1451	44	0	1	3	60	2	8	1569
	Foundation	138	776	–	–	–	–	37	–	1	952

Data from: Nagaoka K, Inazu R, Tohgishi Y, Tanikawa H, Hashimoto S. Estimation of surface/subsurface material stock related to the construction sector of all prefectures in Japan. Sel Pap Environ Syst Res 2009;37:213–19 [in Japanese]; Tanikawa H, Hashimoto S. Urban stock over time: spatial material stock analysis using 4D-GIS. Build Res Inf 2009;37(5–6):483–502.

Table 3.2.2 Material intensities of road classified by structure (unit: kg/m²)

Paving type	Road width	Surface		Base layer		Roadbed	Total
		Asphalt	Aggregate	Asphalt	Aggregate		
Simple asphalt paving	Width < 5.5 m	3.1	43.9	–	–	311.8	358.8
Asphalt paving	5.5 m ≤ width < 13 m	7.6	109.9	7.6	109.9	926.1	1161.1
	13 m ≤ width < 19.5 m	7.6	109.9	6.5	94.0	1144.1	1362.1
	Width ≥ 19.5 m	7.6	109.9	6.5	94.0	1518.1	1736.1
	Highway	7.6	109.9	6.5	94.0	1770.1	1988.1

Data from: Nagaoka K, Inazu R, Tohgishi Y, Tanikawa H, Hashimoto S. Estimation of surface/subsurface material stock related to the construction sector of all prefectures in Japan. Sel Pap Environ Syst Res 2009;37:213–19 [in Japanese]; Tanikawa H, Hashimoto S. Urban stock over time: spatial material stock analysis using 4D–GIS. Build Res Inf 2009;37(5–6):483–502.

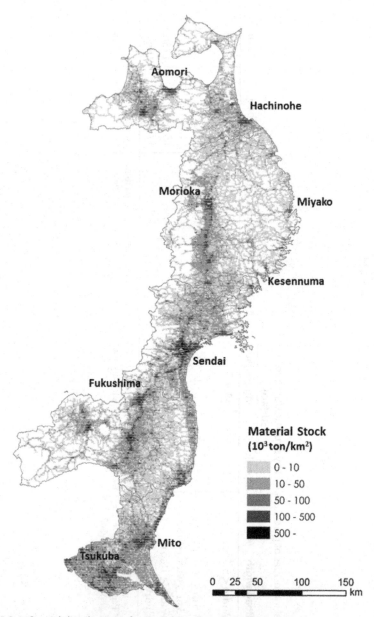

Fig. 3.2.1 Spatial distribution of material stock in the Tohoku region.

3.2.3 AMOUNT OF STOCK LOST DUE TO THE TSUNAMI OF THE GREAT EAST JAPAN EARTHQUAKE

By overlaying the data of tsunami run-up caused by the Great East Japan Earthquake on the spatial information of material stock established in the preceding section, the amount of materials affected is determined, based on which the amount of lost stock is estimated. The target area is the five prefectures of Aomori, Iwate, Miyagi, Fukushima, and Ibaraki, which were seriously damaged by the tsunami. As the data to determine the tsunami range, Maps of the Area Hit by the Tsunami of 11 March 2011 developed by the Association of Japanese Geographers is used. These maps are the 1:25,000 topographical maps showing the range that suffered serious damage, such as sweeping away of houses, and the range of tsunami run-up, which were determined by stereoscopic viewing of aerial photos taken after the earthquake.

Table 3.2.3 shows the result of counting the number of buildings and the amount of materials contained in the affected ranges (1. the range in which most houses were swept away, 2. the range of tsunami run-up area) for each prefecture. For comparison, the status of damage released by Fire and Disaster Management Agency of the Ministry of Internal Affairs and Communications, Japan (FDMA) (Report No. 149; Mar. 7, 2014 version), and the status of progress on disposal of disaster waste in the coastal municipalities (Apr. 25, 2014 version) released by Ministry of Environment, Japan (MOE) are also presented. Table 3.2.4 shows the amount of materials of the roads contained in the affected ranges. We find from the comparison that the number of completely destroyed buildings released by FDMA and the number of buildings in the areas in which most houses were swept away are quite close. As to the amount of rubble, the values of our estimation are smaller for most municipalities, which is probably because only damage from tsunamis is considered in our estimation. In severely damaged areas, such as Rikuzentakata, Kamaishi, Kesennuma, and Minamisanriku, however, our estimated values are larger. Thus, more detailed examination is required.

As a result of the estimation based on the inundated areas released by the Association for Japanese Geographers, the total amount of lost stock of buildings in Aomori, Iwate, Miyagi, Fukushima, and Ibaraki prefectures is found to be approx. 31.8 million ton. The total amount of lost stock of roads in Iwate, Miyagi, and Fukushima prefectures is approx. 2.1 million ton. Looking at the lost stock by the material type, for the upper part of buildings, concrete is approx. 2,269,000 ton, mortar is approx. 1,459,000 ton,

Table 3.2.3 Estimated lost material stock of buildings

	Range in which most houses were swept away				Range in tsunami run-up area					FDMA	MOE
	Number of buildings	Material stock remaining			Number of buildings	Material stock remaining				Completely destroyed (number)	Disaster waste (1000 ton)
		Building	Foundation (1000 ton)	Total		Building	Foundation (1000 ton)	Total			
Aomori	0	0	0	0	4215	514	1220	1734		308	0
Iwate	28,032	814	1530	2344	53,840	795	3577	5372		19,107	5837
Miyagi	45,472	1283	2657	3940	158,847	6487	13,187	19,675		82,911	18,692
Fukushima	8,387	201	404	605	30,196	1160	2480	3649		21,235	3837
Ibaraki	81,891	2298	4592	6890	255,702	10,367	21,440	31,807		126,189	28,366
Total	163,782	4596	9183	13,779	502,800	19,323	41,904	62,237		249,750	56,732

Note: FDMA: Fire and Disaster Management Agency of the Ministry of Internal Affairs and Communications, Japan; MOE: Ministry of Environment, Japan.

Table 3.2.4 Estimated lost material stock of roads

	Range in which most houses were swept away		Range in tsunami run-up area		Regulated road reported by FDMA	
	Materials (1000 ton)	Length (km)	Materials (1000 ton)	Length (km)	Materials (1000 ton)	Length (km)
Iwate	865	201,684	2518	580,676	202	23,060
Miyagi	1154	332,384	10,013	2,882,856	1785	207,072
Fukushima	2190	582,272	15,927	4,376,978	2053	237,221

Note: FDMA: Fire and Disaster Management Agency of the Ministry of Internal Affairs and Communications, Japan.

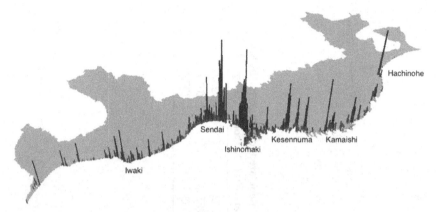

Fig. 3.2.2 Results of estimation of lost stock due to tsunamis.

wood is approx. 2,089,000 ton, glass is approx. 123,000 ton, ceramics are approx. 1,165,000 ton, iron is 2,165,000 ton, aluminum is approx. 60,000 ton, and other materials are approx. 1,038,000 ton. For the bases of buildings, gravel/stone is approx. 6,162,000 ton, concrete is approx. 14,954,000 ton, iron is 313,000 ton, and others are approx. 1000 ton. Fig. 3.2.2 is an overhead view of the distribution of lost stock along the Sanriku Coast.

To avoid misunderstanding regarding these estimation results, it should be noted that the weights presented as the final result according to the definition include materials that have lost their service functions but not been physically swept away by tsunamis, and are therefore not equal to the amount of rubble. Also, since the estimation concerning the structures of buildings or infrastructure is conducted partly based on the assumptions according to relevant documents or on-site surveys, data may be updated as necessary.

3.2.4 MAP LAYERED JAPAN, A WEBSITE TO TRANSMIT INFORMATION

In recent years, serious natural disasters, such as the Great East Japan Earthquake and torrential rains, have raised public awareness of disaster prevention/mitigation. Municipal governments provide hazard maps and information on evacuation facilities through their websites, etc. It is necessary for each individual to think in advance what action to take and where to evacuate if a disaster occurs. To do this, it is also important to obtain information concerning the degree of safety of each local area, as well as the estimated damage, including "lost stock," which has been estimated in previous sections and will contribute to efficient reconstruction of the affected area.

As a technological means to release these various kinds of information effectively, GIS (Geographic Information System) is widely used. GIS is a system to produce, store, apply, and control various maps attached with various relevant data, which can be displayed and searched for reference to geographical information. If you search for a specific store on a conventional paper map, for example, you need to find the name and address of the store before spotting it on the map. By using a GIS map which is combined with the name, address, phone number, and other relevant information of the store, you can display the location of the store by knowing only the name of the store. For the increasing number of users of mobile terminals, such as smartphones and tablet PCs, GIS, when used simultaneously with their GPS function, enables the users to identify the location of a specific spot, as well as to search and display its peripheral facilities. Besides being convenient, these mobile terminals will serve as important communication devices in the event of a disaster, and therefore their access to GIS and its information is made available both at normal times and in an emergency.

Meanwhile, useful information concerning social and economic affairs, the natural environment, and daily life released by municipalities are provided in different file formats or unintegrated coordinate systems, making it difficult to understand the relationships among them. By integrating the format of provision to enable overlaid display of multiple graphic data, relationships among the data can be easily identified, which may lead to the discovery of a new relationship.

In these circumstances, we have organized various spatial information in an integrated format and developed "Map Layered Japan," a website capable of multiple-layered display of such data (Fig. 3.2.3). In Map Layered Japan, developed based on an advanced study in Wakayama, Japan, files of geographical information are integrated into the KML (Keyhole Markup Language) format to enable access through Web browsers of PC or mobile terminals, as well as downloading and displaying on Google Earth. This enables users to visually see the data overlaid with other released information or the data the users have.

In Map Layered Japan, users can choose a map to display from the select box placed in the upper part of the screen and then choose another map data to display on top of it. Also, clicking on a point on the map displays the attribute information of the point. Because the information tabs remain displayed until the "hide" button is pressed, several data can be viewed at a time. Moreover, when viewed in a GPS environment, Google Geolocation API displays the latitude, longitude, and altitude of the current location as

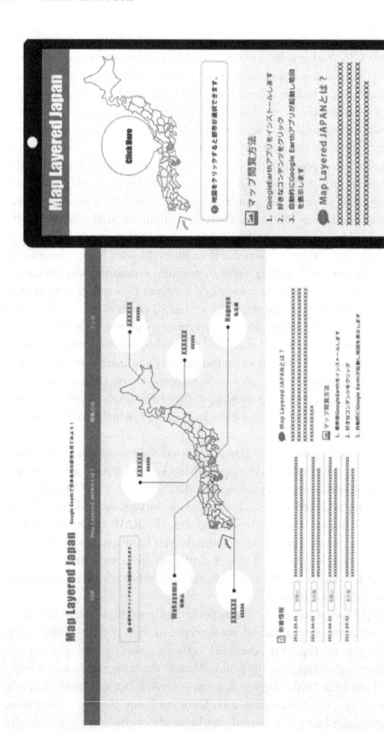

Fig. 3.2.3 Map Layered Japan website (http://danso.env.nagoya-u.ac.jp/maplayered/).

icons on the map. This means that the viewers via a mobile terminal can easily obtain the information on their current location, helping them to determine the route to the nearest evacuation facility and predict the situation of their current location.

Thus, Map Layered Japan will be able to provide visual information throughout Japan if sufficient geographical data are gathered and organized in an integrated format. The data stored in Map Layered Japan, including data on the amount of stock lost by tsunamis, which is estimated in this chapter, are currently being developed. We plan to conduct hearing surveys to verify the effectiveness and convenience of the system and further enhance its functions to satisfy the interests of users.

REFERENCES

[1] Tanikawa H, Hashimoto S. Urban stock over time: spatial material stock analysis using 4D-GIS. Build Res Inf 2009;37(5–6):483–502.
[2] Nagaoka K, Inazu R, Tohgishi Y, Tanikawa H, Hashimoto S. Estimation of surface/subsurface material stock related to the construction sector of all prefectures in Japan. Sel Pap Environ Syst Res 2009;37:213–9 [in Japanese].
[3] Tanikawa H, Managi S, Lwin CM. Estimates of lost material stock of building and roads due to the Great East Japan Earthquake and tsunami. J Ind Ecol 2014;18(3):421–31.

CHAPTER 3.3

Measuring Damage and Recovery Status of Residents in Terms of "Quality of Life"

H. Kato*, Y. Hayashi*, A. Mimuro†
*Graduate School of Environmental Studies, Nagoya University, Nagoya, Japan
†(former) Graduate School of Environmental Studies, Nagoya University, Nagoya, Japan

3.3.1 INTRODUCTION

A disaster not only directly kills or injures people, but also forces those who have survived uninjured to live in an environment that is much worse than before for a certain period of time. This is because their houses, or the infrastructure or facilities for their living, are destroyed or made unavailable, and various services necessary to support their daily life become disabled. This may cause a shortage of foodstuff and other fundamental goods or places, and may hinder rescue, evacuation, restoration, and other necessary activities. Being placed in a bad living environment for a long time may result in impairment of health or even death.

Thus, in discussing advance measures against disasters, besides reducing the number of the dead or injured, measures to secure the living environment of the survivors—in other words, measures to secure redundancy in traffic, electricity, and communication networks and establish places for evacuation and storage—must be considered. Moreover, for districts that are judged to be associated with a long-term threat to the living environment, restriction, or prohibition of locating such facilities may be necessary.

The author summarized the chronological changes in major needs of the affected people in Table 3.3.1 based on Japanese newspaper articles after the Great East Japan Earthquake and other materials including the results of questionnaire surveys of people affected by large-scale disasters. Needs immediately after a disaster are mostly directly relevant to supporting life, such as for food. Needs related to sanitation later increased, while the focus on the need for goods gradually shifted from quantity to quality. Thus, the needs of the affected people change as time passes, while it is difficult to

Table 3.3.1 Major changes in needs of sufferers

Stage	Actualized needs	Explanation
Just after disaster generation	Access to emergency evacuation facilities and medical facilities, food drinkable water, warm shelter; confirmation of family's safety	For preserving life
Evacuation (short-term)	Adequate supply of shelter, water, food, medicine, etc.; sanitary toilet, bathing, clothes, housewares; safety against aftershocks	For a sanitary environment
Evacuation (long-term)	Warm food, privacy, improved sanitary condition, control of infectious disease, air conditioning, recreation	For a healthful and sanitary environment
Restoration	Moving into temporary housing, transport mode to go to reopened companies and schools	For social life
Rehabilitation	Convenience of commuting, going to school, hospital, and shopping; comfort of housing and surrounding environment; preparation for the next disaster	For improving livability

obtain, organize, select, and communicate information on such changing needs, and this is likely to cause confusion. To proceed efficiently with rescue activities under such conditions, advance preparations and prompt follow-up systems are required to allocate limited personnel and goods appropriately and support early recovery to normal social life. As preparation to this end, it is necessary to establish a system to identify and monitor the status of damage and the associated changes in the survival and living environment of the affected people in a chronological order from occurrence of a disaster to restoration from it.

We therefore developed a system with which the survival and living environment of the affected people after a large-scale disaster can be chronologically evaluated in units of small districts by employing the QOL (quality of life) standards, and applied this system to conduct evaluation for 2 months after the occurrence of the Great East Japan Earthquake. Details of this evaluation are as follows.

3.3.2 THE LIVING ENVIRONMENT (QOL) AT THE TIME OF A DISASTER

3.3.2.1 Definition of QOL Standards

Here, the QOL standards at the time of a disaster are defined as "the range of opportunities secured for people's activities (the range of choice)." Identifying and organizing the various determinant factors for this definition are expected to lead to development of an evaluation system.

Togawa et al. [1], considering daily life as repeated cycles of generating needs and satisfying them, focused on the length of each time cycle and categorized the needs related to survival/living environment into four levels, below in order from the most basic needs.

Level 1: What is needed to sustain life (meals, sleep, etc.: in hourly to daily cycles);

Level 2: What is needed to maintain health and sanitation (bathing, washing, etc.: in daily to weekly cycles);

Level 3: What is required for a social being (labor, education, etc.: in daily to weekly cycles);

Level 4: What is needed to maintain cultural life (entertainment, mental rest, etc.: in weekly to monthly cycles).

These show that the shorter the time cycle is, the more crucial to maintenance of life, which is consistent with the changes in the needs of the disaster-affected people shown in Table 3.3.1. When the infrastructure and facilities that satisfy various needs in normal times stop functioning due to a disaster, the needs that arise in short time cycles and are thus crucial to maintaining life seem to supersede all the other needs. This structure is similar to Maslow's [2] hierarchy of needs, according to which people become motivated to achieve the needs in higher stages after their lower-level needs are satisfied.

Based on Table 3.3.1, the relationship between the QOL elements and the QOL standards at the time of a disaster is organized as presented in Table 3.3.2.

3.3.2.2 QOL Standards Evaluation System

Fig. 3.3.1 shows an overview of the disaster QOL evaluation system that the author has developed. In units of small districts (the unit of the Basic Grid Square (approx. 1 km) is employed in the subsequent analysis), whether the needs of the affected people (demand side) are satisfied is judged based on the status of infrastructure and services (supply side). Specific steps for this

Table 3.3.2 Elements constituting QOL

QOL level	QOL components		
Maintaining cultural life (Level 4)	Cultural opportunity	Living comfort	Safety and security
Maintaining social life (Level 3)	Education	Housing	Aftershock risk
	Working	Assurance of rivacy	
	Shopping	Telecommunication	
Maintaining health and sanitation (Level 2)	Availability of bath and shower		Aftershock risk
	Clean clothes		Injury risk
	Medical care opportunity	Privacy	Crime risk
	Foods (quality)	Sanitation level of toilet	
	Life Ware	Air environment	
		Heat environment	
Maintaining life (Level 1)	Drinkable water	Availability of shelter to sleep well	Aftershock risk
	Rescue and medical care opportunity	Air conditioning of shelter in winter and mid summer	Information of family's safety
	Availability of medicine		
	Foods (quantity)		

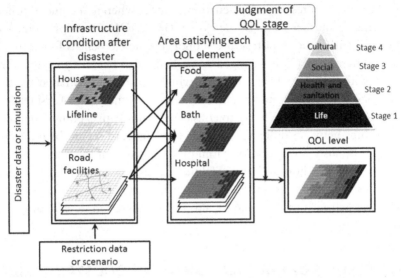

Fig. 3.3.1 Flowchart of a system evaluating QOL level in a disaster situation.

evaluation are: (1) to clarify the status of outage/damage of infrastructure and buildings in the small district units based on the actual data or simulation of damage and data on subsequent restoration; and then based on the status found, (2) to determine the districts where each of the QOL determinant elements (needs of the affected people) can be satisfied; and finally, based on the level of sufficiency determined for each element, (3) to judge for each small district unit which stage of the needs of QOL shown in Table 3.3.2 are satisfied, starting from the bottom, and finalize the QOL level of the district unit. Here, a district of "Level *n*" means a district that is in need of infrastructure and services of Level *n* (not satisfied). A level is considered as being satisfied when all the elements of the level are judged as satisfied, and then moves up to the next level. For example, in a district that is severely damaged by a disaster, where people suffer shortage of food and are forced to live in cold evacuation centers immediately after occurrence of the disaster, these basic needs must be satisfied first, and therefore the QOL standard of this district is judged as Level 1 even if any of the QOL elements of higher levels are satisfied.

In judging the status of satisfaction of each QOL element, an element is judged as satisfied when all the infrastructure and other factors necessary for that element to reach a satisfactory level are fully provided, and as unsatisfied if one of those factors is not fulfilled. Here, it should be noted that a QOL element may be judged as satisfied in two ways: (1) when it is satisfied within the place of residence (evacuation); and (2) when it becomes satisfied by moving to another place. In the case of (1), a QOL element is judged as satisfied when all the infrastructure supporting the element is functioning, while in the case of (2), it is judged as satisfied when necessary facilities are located within the reachable distance, which is determined based on the conditions of roads and public transportation, from the place of residence.

Moreover, it is also necessary to consider the alternative/complementary relationships between different needs. For example, to secure bathing opportunities (satisfy the needs for bathing), at least one of the combinations of needs of either "housing" and "lifeline utilities (water and sewerage, gas) or "means of transportation" and "bathing facilities" must be functioning normally. Such relationships should also be considered in determining the QOL standards.

3.3.3 RESULTS OF CALCULATION OF THE QOL STANDARDS

Using the method described above, we evaluated the QOL standards of Iwate and Miyagi prefectures in the aftermath of the Great East Japan Earthquake. In addition to evaluating the actual case as presented in (1) below, we also conducted an evaluation of case (2), which assumes that the Sanriku Expressway was entirely in service, though it was actually only partially available. Based on a comparison of the results of these two cases, we analyzed how the difference in the status of infrastructure influences the QOL standards of the affected people.

3.3.3.1 Actual Case (Sanriku Expressway Partially in Service)

Fig. 3.3.2 shows changes in the QOL standards in each district from Mar. 14, 3 days after the earthquake, to May 11, 2 months after the earthquake. Because employment opportunities could not be considered (therefore, this element was set as satisfied for all districts) in this analysis, many districts that satisfied all elements of the stage of maintaining health and sanitation (Level 2) turned out to be also satisfying the elements of the stage of maintaining social life (Level 3) other than job employment opportunities at the same time, jumping up to Level 4 from Level 2. As a result, there are very few districts in Level 3.

Overall, inland areas recovered quickly while coastal areas, where infrastructure was severely damaged by tsunamis, took longer to recover.

In coastal areas, as districts that were connected to roads from inland areas were able to secure infrastructure necessary to satisfy the basic needs of

Barometer of Resilience:
QOL Transition after Great East Japan Earthquake

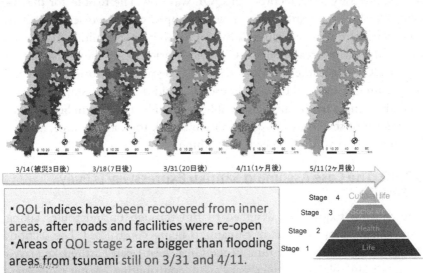

Fig. 3.3.2 Time-serial change in QOL level.

the affected people, the QOL standard level rose from Level 1 to Level 2 on the third day from the earthquake. This seems to be the effect of operation "Teeth of a Comb" (conducted jointly by the prefectural governments and the Japan Self Defense Forces to secure routes for emergency transportation in the shape of teeth of a comb by eliminating obstacles and clearing roads).

As of Apr. 11, 1 month after the earthquake, many districts in coastal areas had recovered to Level 2. These districts, however, are located in a range of areas larger than the tsunami inundation areas. This indicates that movement of the residents of upland areas, which are not included in the tsunami inundation areas, were substantially restricted despite little direct damage to their houses because their important facilities were swept away and major roads were damaged by tsunamis. In addition, suppliers of lifeline utilities, such as water treatment plants, were also affected by the earthquake, causing a significant impact on the entire area covered by them. As of May 11, 1 month later, most districts had not moved up to higher levels from Level 2. This is because it was taking a long time to restore the portions of infrastructure damaged by tsunamis.

In inland areas, which suffered no damage from tsunamis, convenient facilities recovered quickly and road networks were sufficiently available, therefore the QOL recovered quite early. However, in Kurihara City, an inland city of Iwate Prefecture which marked intensity 7 on the Japanese seismic intensity scale, recovery of QOL was slow. The reasons for this include the low restoration rate of water supply, which worsened sanitary conditions of toilets and caused a lack of bathing opportunities. A comparison of the inland area of Sendai City and Morioka City shows that recovery of QOL was earlier in Morioka. This is because in Morioka, liquefied petroleum gas was used by most households and therefore restoration work could be conducted on an individual household basis, while in Sendai City, restoration of the entire city gas piping network took a long time, making it impossible to satisfy the needs for bathing.

3.3.3.2 The Case Assuming Sanriku Expressway is Fully in Service

The Sanriku Expressway, designed to run through upland routes taking into consideration the risk of tsunami, is being constructed. The sections that were in service at the time of the Great East Japan Earthquake played an important role as an emergency road, contributing greatly to evacuation of residents and restoration work. Thus, we analyzed the impact of this expressway on changes in QOL, by assuming that all the planned sections (Fig. 3.3.3) were in service and suffered no damage.

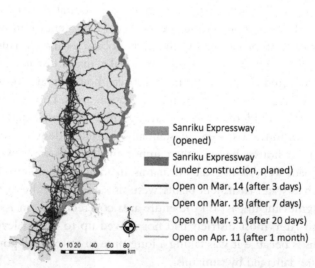

Fig. 3.3.3 Sanriku Expressway.

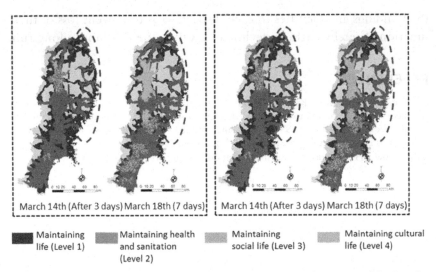

March 14th (After 3 days) March 18th (7 days) | March 14th (After 3 days) March 18th (7 days)

Maintaining life (Level 1)   Maintaining health and sanitation (Level 2)   Maintaining social life (Level 3)   Maintaining cultural life (Level 4)

Fig. 3.3.4 Time-serial change in QOL level. *Left*: real case (partial opening of the Sanriku Expressway); *right*: fully open.

Fig. 3.3.4 shows the changes (in a week after the earthquake) in QOL standards of each district for both the actual case (Sanriku Expressway is partially in service) and the full-service case. In the actual case, the QOL of coastal areas is mostly Level 1 as of Mar. 14 while in the full-service case, the QOL of the areas along the Sanriku Expressway, as well as the areas along the roads that are horizontally diverting therefrom, have recovered to Level 2. Moreover, the numbers of affected people (counted on the night-time population basis) of coastal municipalities in each QOL Level presented in Fig. 3.3.4 indicate that the number of affected people in Level 1 is approx. 8.5% lower (approx. 120,000 people) in the full-service case as of Mar. 14. However, the difference between the two cases had not widened a week after the earthquake. This is probably because even though a broad road network is available, local roads, houses, lifeline utilities, and convenient facilities inside each district—which are supposed to provide services that cannot be obtained in other districts accessible by the road network—are destroyed. This means that lower levels of QOL elements cannot be satisfied, making it difficult for the residents to live an independent life within the local area.

The examination above thus indicates that strengthening of road networks is effective in securing QOL immediately after a disaster, while measures should be implemented to enhance safety against disasters of the buildings and facilities in each district, so as to prevent loss of their functions, thereby curbing decline of QOL in the medium to long run.

REFERENCES

[1] Togawa T, Mimuro A, Kato H, Hayashi Y, Nishino S, Takano T. Evaluation, post-disaster reconstruction and improvement management from QOL standards in disasters. In: The proceedings of the IESL-SSMS joint international symposium on social management systems 2011, Colombo, Sri Lanka. 2011; p. 362–9.
[2] Maslow AH. A theory of human motivation. Psychol Rev 1943;50(4):370–96.

CHAPTER 4

Regional Grand Design for Improving Disaster Resilience

KEY MESSAGES

- Based on detailed analyses of the 2011 Tohoku Tsunami, a new concept of tsunami disaster mitigation was established by introducing two levels of tsunami hazard. Seawalls will thereafter be designed on the basis of the level 1 tsunami. In this section, a new framework is presented for the optimal design of the seawall height. It is based on the estimation of the residual risk. It is demonstrated that the new framework will determine the optimal seawall height by accounting for the risk reduction due to the land use regulation and evacuation measures.
- Chapter 4.2 introduces the development flow of a database for evaluation of initial risk and readiness of large-scale earthquakes on 60 million buildings all over Japan to integrate various census data and building scale micro geo data, ie, the "Micro Geo Data (MGD)." In addition, we developed the data environment to be able to estimate damage situations by earthquakes in various regions and to compare among them

scale-seamlessly by calculation of initial risk and readiness quantitatively to integrate them with seismic intensity. It is expected that our result can make great contributions in policy planning for realization of "resilient" regional grand design.

- Community-based support—action taken at the level of communities—plays an important role in encouraging the development of resilient urban communities. Yet gaining consensus on the action to take is difficult, and firmly establishing a feasible system of community-based support is a challenge faced by society. In addition, estimates of anticipated damage conventionally released by the government have presented an average scenario, which is insufficient to foresee disasters in detail, understand what measures to take, and build social consensus. To meet these challenges, by applying recent simulations that use various kinds of detailed information collected as geo big data, it is important to examine more detailed disaster scenarios and share them among those who carry out simulations.

- It is necessary to apply a technique for quantitative analysis of sustainability and resilience that can examine details of spatial conditions resulting from maintenance of urban and residential areas and infrastructures, in order to discuss how national-level land planning and urban planning should address the occurrence of huge earthquakes and development of climate change. This chapter proposes a methodology to quantify resilience against huge disasters through estimation of a decrease of Quality of Life (QOL) over time. By viewing the factors that threaten good health and life, in addition to the decline in the standard of living, as a reduction in life years of residents, it has become possible to assess comprehensively not only various factors that decrease QOL, but also measures and programs against disasters.

- Extreme floods are happening all over the world in recent years due to climate change. Conventional flood protection measures such as constructing levees and flood control facilities have become inadequate for protecting life and assets against extreme floods. The Netherlands and Japan, both flood-prone countries, are now introducing new adaptive measures with land-use management.

- Despite economic, scientific, and technological advances, damage and losses by natural disasters still remain at a high level in Japan. This is mainly because many people still reside in disaster-prone places both in urban and rural areas. We propose some policy measures to promote relocation from vulnerable places to safe places using financial incentives to residents and local governments.

CHAPTER 4.1

Optimal Seawall Height Based on Risk Analysis with Land Use and Soft Measures

S. Sato
Department of Civil Engineering, The University of Tokyo, Tokyo, Japan

4.1.1 STRUCTURAL (HARD) COUNTERMEASURES AND NONSTRUCTURAL (SOFT) COUNTERMEASURES IN TSUNAMI MITIGATION STRATEGY

Taking precautions against tsunamis is a major challenge faced by Japan, which has suffered many tsunami disasters and will continue to do so, when the country tries to enhance resilience of its coastal areas where the country's population and assets have gathered. It is difficult to predict in detail how high tsunamis will be because their height greatly depends not only on the magnitude and characteristics of earthquakes, but also on the geography of the seabed or coast. Tsunamis happen less frequently than floods and run-up of storm waves, and compared to earthquakes, which cause damage instantly, it takes more time for tsunamis to reach and hit the coast after they occur off the coast. Tsunami disaster mitigation has been implemented by combining structural (hard) countermeasures and nonstructural (soft) countermeasures. The structure-based countermeasures are represented by seawalls and tsunami breakwaters. The nonstructure-based countermeasures include relocation of residential area to higher places, early warning, and evacuation.

Estimating the scale of storm waves, storm surges, and tsunamis that will reach coastal areas requires study on how waves develop in the spatial scale of tens of kilometers. The Seacoast Law requests, therefore, that prefectural governors work with the central government to take measures for coastal disaster prevention. For example, in Japan the height and structure of seawalls, which are an example of structural measures, are determined by prefectural governors in their role as coastal managers. Before these are determined, not only the highest water elevation of tides, storm surges, and run-up of storm waves, but also the design tsunami height are taken into consideration to determine the specifications of seawalls, so that seawalls will be high enough to prevent sea water that is as high as the highest water elevation and design tsunami height from going over the walls to the land.

On the other hand, in order to consider what precautions should be provided against other disasters, such as fire and earthquakes, it is more practical to take account of conditions of nature and society that are specific to communities and districts; therefore, such precautions are considered at a smaller level, like cities, towns, and villages, than that of coastal management, in accordance with the Basic Act on Disaster Control Measures, which is a basic law for provision of such precautions. The tsunami disaster mitigation strategy is drawn up as part of local disaster prevention plans which are arranged by mayors of cities, towns, or villages, and is aimed at providing an evacuation program in the event of high tsunamis that cannot be blocked by seawalls. Even land areas protected by seawalls cannot be perfectly protected, and buildings and houses there will be flooded if tsunamis flow over the seawalls into those areas. However, it often takes some time for a tsunami to reach the coast after an earthquake has occurred, so quick evacuation can save more human lives. Nonstructural measures aimed at disaster reduction are taken to prevent loss of human lives and reduce damage to assets, and are different from structural measures, which are taken to prevent floods and are aimed at disaster prevention. To implement disaster reduction based on quick and well-organized evacuation, it is important for each resident in a community to maintain awareness of disaster prevention and for the residents of the community to help each other. It is important, therefore, to understand the role and limitations of seawalls, which are recognized to be public support, and translate the concept of community support and individual support into detailed action plans. It was also pointed out even before the Great East Japan Earthquake how important public support, community support, and individual support were, because the experiences of the 1993 Okushiri Tsunami, the 2004 Indian Ocean Tsunami, the 2005 Storm Surge due to Hurricane Katrina, etc. had indicated that tsunamis and storm surges caused by these disasters were too large to be prevented by seawalls alone.

Fig. 4.1.1 is a schematic diagram and represents a concept of a comprehensive disaster prevention program that combines structural and nonstructural measures [1]. The horizontal axis represents the scale (height) of a tsunami, while the extent of damage is represented in the negative direction on the vertical axes. Because the extent of damage caused by a tsunami accelerates as the tsunami becomes higher, the damage when

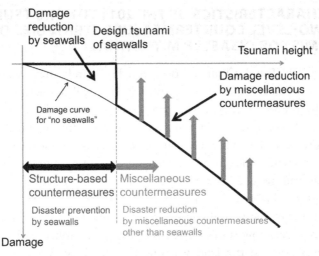

Fig. 4.1.1 Tsunami disaster prevention based on structural (hard) countermeasures and mitigation based on nonstructural (soft) countermeasures (before 2011).

no measures are taken to prevent the tsunami will become increasingly serious as the tsunami develops high waves, and the extent of the damage is represented by a convex curve that goes up from the left and then down to the right. To take structural measures that provide constructions like seawalls, the scale of the tsunami to be blocked by the measures is determined based on data that includes the record of the largest recorded tsunami, as shown in the figure, and examination of how the run-up of storm waves or storm surges will behave. Then, constructions, such as seawalls that are designed based on the previously mentioned information, will be set up along the coast to prevent tsunamis from flooding the land. Seawalls, however, are designed against an external force that is not large enough to go over the levee crown of the walls; the walls cannot be expected to be effective against and safe from a tsunami that is higher than they are. It is the philosophy of the comprehensive tsunami mitigation that, in the event that water flows over the seawalls and causes a flood, nonstructural measures with a focus on a quick evacuation should be taken to minimize the damage caused by the tsunami. Structural measures include constructing seawalls to prevent the land from being flooded, and these deserve to be called disaster prevention, while nonstructural measures should be regarded as disaster mitigation because they are aimed at saving more human lives and reducing the extent of damage when urban areas are flooded.

4.1.2 CHARACTERISTICS OF THE 2011 TOHOKU TSUNAMI AND TWO-LEVEL COUNTERMEASURE STRATEGIES OF TSUNAMIS FOR DISASTER MITIGATION

When the 2011 Tohoku Tsunami occurred, Fudai and other areas in northern Iwate Prefecture were protected by seawalls as high as the tsunami, which successfully prevented the tsunami from coming to the land. However, in northern Fukushima Prefecture and in many areas further north, the tsunami was several meters higher than the seawalls there and wrecked them. Even in areas where the seawalls survived, it was difficult to confirm clearly that the seawalls reduced the damage of the tsunami. In contrast, in southern Fukushima Prefecture and areas further south, the overflow depth of the tsunami over the seawalls was approximately 1–5 m, and a clear relationship between how seriously the seawalls were broken and how much damage was done to the land was observed in these areas. For example, some seawalls along the Nakoso Coast in Fukushima Prefecture were hit by a tsunami with approximately 1-m overflow depth, but were not destroyed, resulting in minor damage to the land area. Most walls in another area a few 100 m away that were hit by the tsunami with an approximately 3-m overflow depth were ruined and could not prevent a large-scale flood [2]. In addition, many coastal areas of Minami-Soma City suffered large-scale flood damage, but the extent of the damage tended to be smaller in areas with a lower percentage of totally destroyed seawalls; it was also confirmed by this fact that seawalls, if not totally broken, effectively reduced the amount of overflow [3].

In addition, the fact that the Tohoku Tsunami flooded some evacuation places led to a recognition of how important it is to set in a detailed and scientific manner the scale of the tsunami allowed for to plan structural and nonstructural measures. Although the tsunami, which was far larger than the set conditions, tore down many seawalls, the walls that were not destroyed completely are reported to have lessened the flood damage caused by the tsunami, as observed in the examples of Nakoso and Minami-Soma cities, and the effectiveness and limitations of structural measures will be quantitatively explained soon. Based on these findings, an idea is being introduced to set the scale of tsunamis in two levels, as shown in Fig. 4.1.2, level 1 tsunamis are those that occur once in a few decades to a hundred years plus several decades, and are used to design constructions like seawalls, and level 2 tsunamis are those that occur less frequently, or once in more than several 100 years, and are used to draw up evacuation plans, etc. In addition,

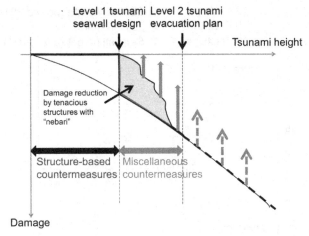

Fig. 4.1.2 Tsunami disaster mitigation based on level 1 tsunami and level 2 tsunami (after 2011).

a tenacious structure with a "nebari" concept that is also effective against tsunamis exceeding its design height is also being considered. Every effort is being made to combine detailed, practical nonstructural measures and resilient structural measures to save human lives and reduce damage to assets.

4.1.3 HEIGHT OF SEAWALLS IN IWATE PREFECTURE

Comprehensive preparations against tsunamis had been provided to Japan's eastern coast, which was hit by the giant Tohoku Tsunami in 2011. Fig. 4.1.3A compares the heights of tsunamis near the coast that struck Iwate Prefecture and of seawalls there. The Meiji Sanriku Tsunami (1896), which carried high waves and hit northern Iwate Prefecture in particular, is the reason high seawalls were provided along the northern coast of the prefecture. The Sanriku region has experienced so many high tsunamis referred to in records, which were higher than the run–up height of storm surges and storm waves, that the height of the tsunamis were allowed for to determine the height of the seawalls built in many areas of the Sanriku coast. However, because coasts of Japan are exposed to a severe wave climate, including typhoons and winter storms, coasts with such high seawalls are rather exceptional in the country; it is more common on the coasts in this country that the run–up height of storm surges and storm waves are higher than the height of tsunamis. In the Tohoku region hit by the tsunami, the height of seawalls in Sendai Bay and on the coast of Fukushima Prefecture,

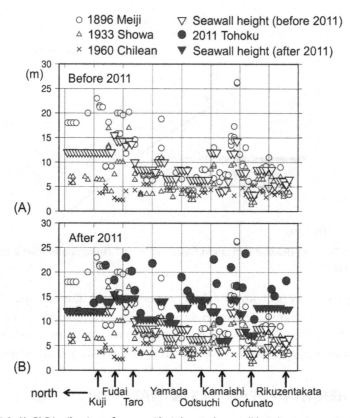

Fig. 4.1.3 (A, B) Distribution of tsunami heights and seawall heights in Iwate Prefecture.

excluding only some exceptions, is determined based on the height of storm surges and storm waves. Also in the example of Iwate Prefecture shown in Fig. 4.1.3A, the largest recorded tsunamis triggered by the 1896 and 1933 Sanriku earthquakes and the 1960 Valdivia earthquake (Chile), respectively, were lower than the run-up height of storm surges when they reached the southern coast of the prefecture; therefore, the height of seawalls was determined based on the height of storm surges. For example, Rikuzentakata City had a pine forest at its coast, which is reported to have reduced a flood in the land behind the forest when the Chilean tsunami occurred in 1960; within the pine forest was a five- to six-meter-high embankment.

Fig. 4.1.3B shows the height of the 2011 tsunami and that of seawalls determined by an analysis conducted after the tsunami. The analysis used the data on the height of tsunami near the coast, published by a joint research group (2011) [4] and Iwate Prefecture (2011) [5]. The 2011 tsunami that hit Rikuzentakata City was much higher than those that had occurred before and broke the pine forest and seawalls almost completely. Rebuilding the seawalls to a height of 12.5 m to prevent flooding caused by a level 1 tsunami is being planned.

4.1.4 OPTIMAL COMBINATION OF VARIOUS TSUNAMI COUNTERMEASURES BASED ON RESIDUAL RISK ANALYSIS

As described above, it has been decided to promote preparations against tsunamis, by setting the two tsunami levels based on their frequency and by combining structural and nonstructural measures. Seawalls are designed against level 1 tsunamis, which occur once in a 100 years or so. This means that only the frequency of tsunamis is considered to decide the height of seawalls. This approach to setting the height of seawalls is reasonable to some extent to secure a common extent of safety, when most coastal low-lying areas are used as places of residence, as in the case of Japan. The extent of safety in communities can be raised by measures other than seawalls, and it is recommended to take account of such conditions to design the height of seawalls. As one of the ways to enhance safety against tsunamis, it is effective to change the way lands are used, through relocation of communities to uplands and regulation of the use of low-lying coastal areas, for example. Provision of education in disaster prevention and evacuation drills to more people will raise the rate of successful evacuation and lead to a reduction of casualties. It is important to have a framework to consider both conditions of nature comprehensively, such as characteristics observed in the event of earthquakes and tsunamis, and social conditions that have an impact on the optimal design height of seawalls.

A residual risk analysis is an effective means of examining such conditions comprehensively. Fujima and Hiwatashi [6] studied Toi district in Shizuoka Prefecture to assess the residual risk of the district and, based on the assessment, proposed a method to determine the optimal height of seawalls. The residual analysis performed to consider preparations against tsunamis is to analyze the damage expected to be caused by, as well as the incidence of, a tsunami in the future, for example in the next 50 years in an area forecasted to be flooded by the tsunami, and calculate the expected value D of the damages in the area.

$$D = \int_0^\infty d\left(H, h_c\right) p\left(H\right) dH$$

Values H and $p(H)$ represent the height of an approaching tsunami off shore and the probability of the tsunami as high as H reaching the area, respectively.

Values h_c and $d(H,h_c)$, respectively, represent the heights of seawalls and financially calculated amounts of the damage expected to be caused by the tsunami. This function anticipates that the amount will be increased as the height of the tsunami H becomes higher and will be decreased as the height of the seawalls h_c becomes higher. The above formula calculates the expected value, which is an amount of the damage anticipated in the area over the period subject to study. The expected value of the damage is not normally zero in our country, where its population and assets gather in flat, low-lying coastal areas, because there is a finite probability that a tsunami will overtop seawalls and reach the coastal areas; seawalls do not free the country from a risk of flood caused by tsunamis. The purpose of various measures taken against tsunamis is, therefore, to reduce the residual risk.

Fig. 4.1.4A is a graph in which heights of seawalls h_c and financial amounts are respectively plotted on the horizontal and vertical axes, and gives an example of the residual risks and construction costs. Residual risks will decrease and construction cost will increase in proportion to the height of seawalls. Since the sum of residual risks and construction costs is the total social cost to be borne over the period by the community, such a height of seawalls that minimizes the total cost will be the most reasonable height. This approach to determining the height of seawalls analyzes not only the incidence of tsunamis, but also the amount of damage that depends on the amount of assets and house types in the community, and is thought to be more rational than the current method, which considers only the incidence of tsunamis to determine the height.

As indicated by Figs. 4.1.1 and 4.1.2, damage caused by tsunamis can be decreased not only by seawalls—structural measures—but also by

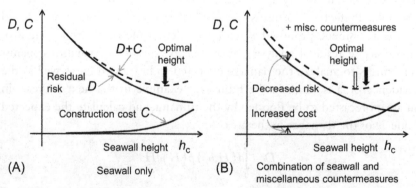

Fig. 4.1.4 (A, B) Optimal height of seawalls based on residual risks and costs for countermeasures.

nonstructural measures, such as relocation of communities to elevated places and faster evacuation. If the amount of the residual risks that can be reduced by taking measures other than the construction of seawalls is quantitatively evaluated, such other measures can be implemented in addition to the construction of seawalls and can minimize the residual risks comprehensively. In this way, it will be possible to determine an optimal combination of measures. For example, according to Fig. 4.1.4B, building seawalls as high as h_c and taking other additional measures will increase cost but further reduce residual risks. If the effectiveness of the measures to take is allowed for and the height of seawalls are determined so that $D + C$ will be minimized, it will be possible to reduce the height of seawalls and mitigate damage.

To analyze which measures to choose to have an optimal combination as explained above, it is necessary to sort out data on the incidence of tsunamis and other basic data of damage, including breakage of seawalls, and changes in the way land is used over the period subject to study. Since conventional design of seawalls did not consider how seawalls, etc. are broken by tsunamis flowing over the walls and how effective the walls are in mitigating damage, there are scarcely any such studies and analyzes available for reference. To promote such measures that will save more people from tsunamis, it is necessary to evaluate not only the ability of seawalls to withstand overtopping, but also to assess a combination of structural and nonstructural measures comprehensively from various perspectives, including those of social equality, economy, and risk management, and carry out the measures smoothly. Reducing residual risks of tsunamis and improving resilience of coastal areas call for a preliminary disaster assessment program aimed at encouraging a certain manner of land use and a comprehensive coastal management system that collectively deal with measures against tsunamis; our upcoming task is to interdisciplinarily discuss and study how to implement such a program and system.

REFERENCES

[1] Sato S. Characteristics of the 2011 Tohoku Tsunami and introduction of two level tsunamis for tsunami disaster mitigation. Proc Jpn Acad Ser B Phys Biol Sci 2015;91:262–72.

[2] Sato S, Takewaka S, Liu H, Nobuoka H. Tsunami damages of Nakoso Coast due to the 2011 Tohoku Tsunami. In: Proceedings of the 33rd International Conference on Coastal Engineering (ICCE), Santander; 2012. p. 320–34.

[3] Sato S, Okayasu A, Yeh H, Fritz HM, Tajima Y, Shimozono T. Delayed survey of the 2011 Tohoku Tsunami in the former exclusion zone in Minami-Soma, Fukushima Prefecture. Pure Appl Geophys 2014;171(12):3229–40. http://dx.doi.org/10.1007/s00024-014-0809-8.

[4] TTJS (Tohoku Tsunami Joint Survey Group). Tsunami survey database, http://www. coastal.jp/tsunami2011; 2011 [accessed Aug. 2015].

[5] Iwate Prefecture. Seawall heights in Iwate Prefecture, http://www.pref.iwate.jp/ kasensabou/kasen/fukkyuu/008326.html; 2011 [accessed Aug. 2015; in Japanese].

[6] Fujima K, Hiwatashi Y. New approach to specify the adequate scale of facility against tsunami and the residual risk. J JSCE A1 2013;69(4):I_345–57 [in Japanese].

CHAPTER 4.2

Earthquake Damage Risk Evaluation by Micro Geo Data

R. Shibasaki*, Y. Akiyama*, Y. Ogawa†
*Center for Spatial Information Science, The University of Tokyo, Chiba, Japan
†Graduate School of Frontier Sciences, The University of Tokyo, Chiba, Japan

4.2.1 ISSUES IN THE EXISTING EARTHQUAKE DAMAGE RISK EVALUATION

There is always a possibility, whether big or small, that a large-scale earthquake may cause widespread damage in Japan. Therefore, local governments in each part in Japan have individually conducted surveys to estimate earthquake damage and danger, and have disclosed the results, providing an opportunity for their residents to see and utilize the results.

However, so far, most of this information has been made and disclosed by each local government separately. The information thus obtained is often disclosed collectively only by the unit of town or street, or by the unit of mesh. Furthermore, the methods and standards of the surveys are not necessarily the same, which makes regional comparisons difficult. Therefore, the information currently available to the public is inadequate for responding to the issues for planning the current disaster-prevention measures such as preparing for widespread damage beyond the boundaries of local governments, and improving the disaster-prevention capability of the residents.

For these issues, by using various statistical and space data, which are available to the public and yet cover all of Japan, we have prepared high resolution geo spatial data (micro geo data) to evaluate earthquake damage risk (the risk is called, more concretely, "building collapse and fire risk" due to earthquake) and the initial capability of responding to the damage [1]. We have also proposed a simplified evaluation method utilizing this data for the evaluation of earthquake damage risk and the initial capability of responding. Making this data available will produce an environment in which we can evaluate and compare quantitatively and precisely the damage risk and initial capability of responding for each region of the entire Japan when a large-scale earthquake occurs.

4.2.2 PREPARING THE MICRO GEO DATA FOR EACH BUILDING

First, we obtained the location information of approximately 60 million buildings in Japan from the digital residential map[1] and then prepared the data (hereafter referred to "building point data") that enables us to observe the distribution of all buildings. This data allows us to observe the locational information, area, number of floors, and building use[2] of each building. By allocating various types of statistical information and micro geo data to this data, we prepared an environment that enables us to calculate the damage risk for each building and the initial capability of responding to the disaster.

The following information is added to each building (Fig. 4.2.1):
- information on earthquake damage risk:
  - the possibility that a fire will break out after an earthquake, burning down the house (fire risk),
  - the possibility that an earthquake tremor will cause the building to collapse (collapse risk);
- information on initial capability of responding to earthquake damage:
  - the capability the neighboring fire-fighting organization to extinguish a burning building (this is called "fire-fighting capability" in this study),
  - the capability of neighbors to rescue people from a collapsed building (this is called "mutual assistance capability" in this study);
- information on human risk:
  - the possibility that residents there may be injured (human risk).

Finally, various information regarding earthquake damage is added to all buildings. By adding that information for any space unit (eg, by town, street, school district, mesh), damage risk and initial capability of responding (the human risk due to building collapse and fire) can be calculated. Also, by subtracting the initial capability of responding from the damage risk, the human risk in the district can be estimated.

Based on the flow of Fig. 4.2.1, we shall briefly introduce in the following how to estimate the damage risk for each building (fire and collapse risk), the initial capability of responding, and the human risk immediately after an earthquake. For more details of this method, please refer to Akiyama et al. [2–4], Kato et al. [5], and Ogawa et al. [6].

---

[1]Zmap TOWN II (Zenrin Co., Ltd) provided by the Joint Research Application System (JoRAS), Center for Spatial Information Science, the University of Tokyo (research ID: 122). The data for 2008–09 was used for our results.

[2]The uses are classified into houses, offices, multi-use buildings, landmark buildings, and others.

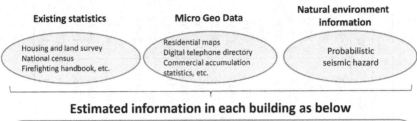

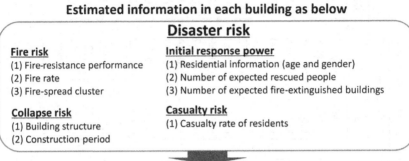

Fig. 4.2.1 Development flow of building micro geo data.

- Calculation of fire risk (①–③)

  ① The estimation of fire-resistance performance

In this method, we supposed that nonwooden structures are either fireproof structures or semifireproof structures. By using the Housing and Land Survey (2008) that lists the number of nonwooden structures as single housing and multifamily housing for each city, ward, town, and village, we found the ratio of wooden structures and nonwooden structures for each city, ward, town, and village. To match this ratio, we allocated for each building the information of either a wooden structure or nonwooden structure. When we allocated, we estimated statistically by combining various information such as the number of floors of each building, judgment result of either "inside" or "outside" of the commercial area,[3] building use, area size, and various other information.

  ② The estimation of fire probability

As shown in Table 4.2.1, the fire probability of a building can be decided based on the building use and seismic intensity. In other words, if we know the building use, we can tell the fire probability of the building. Therefore,

---

[3]The Commercial Accumulation Statistics [2]: the data in these statistics was used to find the distribution of commercial areas and districts.

Table 4.2.1 Fire occurrence ratio of each seismic intensity and building use

Japan Meteorological Agency (JMA) seismic intensity scale

Type of industry	5-Lower		5-Upper		6-Lower		6-Upper		7	
	DS	IW	DS	IW	DS	IW	DS	IW	DS	IW
Theater	0.0043	0.0039	0.0116	0.0125	0.0300	0.0305	0.0832	0.1005	0.1865	0.2956
Cabaret	0.0000	0.0041	0.0000	0.0100	0.0000	0.0242	0.0006	0.0860	0.0229	0.2902
Bar	0.0049	0.0058	0.0044	0.0086	0.0131	0.0231	0.0323	0.0771	0.0954	0.2292
Restaurant	0.0069	0.0073	0.0096	0.0106	0.0291	0.0306	0.0808	0.0858	0.2058	0.2168
Department store	0.0271	0.0211	0.1000	0.0774	0.2531	0.1928	0.7232	0.5694	1.8200	1.6071
Article store	0.0017	0.0014	0.0041	0.0042	0.0107	0.0105	0.0384	0.0458	0.3243	0.3866
Hotel	0.0148	0.0151	0.0644	0.0653	0.1600	0.1618	0.4566	0.4752	0.9663	1.0709
Apartment	0.0007	0.0012	0.0011	0.0027	0.0031	0.0070	0.0090	0.0249	0.0349	0.0757
Hospital	0.0045	0.0035	0.0093	0.0089	0.0247	0.0222	0.0701	0.0759	0.2191	0.4329
Clinic	0.0013	0.0014	0.0013	0.0034	0.0040	0.0082	0.106	0.0282	0.0495	0.1250
Dormitory	0.0014	0.0016	0.0028	0.0025	0.0075	0.0068	0.0228	0.0244	0.1116	0.1456
Nursery school	0.0025	0.0002	0.0033	0.0009	0.0095	0.0019	00.246	0.0094	0.0694	0.0393
Kindergarten	0.0019	0.0013	0.0019	0.0042	0.0056	0.0109	0.0137	0.0594	0.0431	0.1772
Elementary school	0.0083	0.0022	0.0136	0.0058	0.0374	0.0142	0.1002	0.0612	0.2989	0.2175
University	0.0037	0.0007	0.0062	0.0020	0.0170	0.0050	0.0458	0.0155	0.1263	0.0604
Public bath	0.0006	0.0009	0.0009	0.0027	0.0026	00064	0.0073	0.0225	0.0282	0.0874
Factory	0.0016	0.0013	0.0046	0.0046	0.0118	0.0117	0.0330	0.0564	0.0796	0.1529
Office	0.0024	0.0012	0.0069	0.0038	0.0176	0.0095	0.0496	0.0307	0.1208	0.0980
House	0.0007	0.0016	0.0007	0.0035	0.0021	0.0094	0.0058	0.0505	0.0274	0.1521

DS is day time in summer, IW is evening in winter.

in this method, by combining the digital telephone directory[4] with the building point data, we clarified the function of each building and further estimated the fire probability of all buildings by spatially joining the earthquake intensity information.

③ The introduction of fire-spread cluster

As was seen in the Great Hanshin-Awaji Earthquake, there is a chance that a large-scale fire, involving multiple buildings, may break out immediately after a large-scale earthquake. In other words, even if the fire probability of a building is small, we need to consider that the fire risk will become higher due to the spread of fire from surrounding buildings.

Therefore, in our study, we calculated, based on the method of Kato et al. [7], the fire-spread risk of the area where there is a high possibility of fire-spread, which is called a fire-spread cluster, and introduced the fire risk due to fire-spread, thus making it possible to calculate the probability of the building burning down, which reflects the actual situation more closely. The probability of a building burning down is also affected by the fire-spread cluster itself, in addition to the fire probability calculated based on the building use in ②.

By combining the results of the above ①–③, the estimation of the probability of burning down was completed.

- Calculation of building collapse risk (④–⑤)

④ The estimation of building structure

According to the Kobe City survey of the Great Hanshin-Awaji Earthquake in 1995, both the total collapse rate and total/half collapse rate were the highest for wooden structure buildings, while those rates were lower in nonwooden structure buildings such as steel structures, and reinforced concrete structure buildings. Therefore, in this method, we developed a method to estimate whether the building structure is wooden or nonwooden structure.

By using the already prepared building point data with fireproof function, and the Housing and Land Survey (2008) which lists the number of nonwooden structure buildings by single housing and by multifamily housing for each city, ward, town, and village, we estimated

---

[4]Telepoint Pack DB (Zenrin Co., Ltd) provided by the JoRAS, Center for Spatial Information Science, the University of Tokyo. The data for 2010 was used for our results. Since the longitude and latitude information has been added to all telepoint data, the information was used to combine the business category of the telepoint data with the nearest building point data.

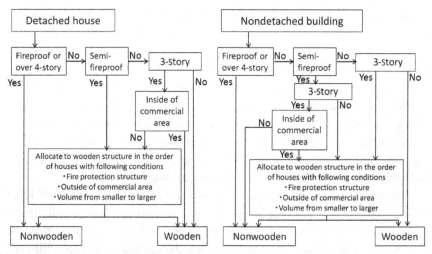

Fig. 4.2.2 Estimation method of building structure.

the building structure according to the method in Fig. 4.2.2. As a result, we found that fireproof buildings, buildings outside the commercial areas, and buildings of small capacity tended to be wooden structured.

⑤ The estimation of construction period

In our evaluation of the building collapse risk, the construction period of a building has a great influence on the collapse risk in addition to the building structure. In particular, the Enforcement Ordinance of the Building Construction Standards Act took effect on Jun. 1, 1981, and new earthquake resistance standards were introduced. Therefore, the resistance to earthquakes differs greatly before and after the act. However, the data of the construction period for each building is available for the public only as an aggregate value for each city, ward, town, and village. In order to reflect the diversity of building collapse probability among the various buildings in the estimation of earthquake damage, it is ideal that the individual building construction period information is available rather than the aggregate value.

Therefore, in our method, we combined each statistical table for building purposes, building structures and numbers of floors that are available from the Housing and Land Survey (2008), and disclosed the ratio of the construction period under various conditions. Based on the ratio, we carried out a micro simulation called the Markov Chain Monte Carlo method and assigned the construction period for each building.

Furthermore, using the Polygon data[5] of the DID (Densely Inhabited District) for the past and present, we made a weighted calculation by the DID-appointed period so that we could reflect the regional characteristics better. In this method, each building was assigned one of five construction periods: (1) pre-1970; (2) 1971–80; (3) 1981–90; (4) 1991–2000; and (5) 2001 up to the present.

By combining the results of the above estimations ④ and ⑤, we could estimate the collapse risk for each building in Japan.

• The initial capability of responding immediately after an earthquake (⑥–⑧)

In this method, we evaluated the initial capability of responding by ⑦ the number of people expected to be rescued from collapsed buildings by neighbors, which is called "mutual assistance capability," and ⑧ the number of buildings where a fire was expected to be extinguished through the public fire-fighting capability such as fire engines, and professional and volunteer fire-fighters, which is called "fire-fighting capability." We also estimated ⑥ the residents' information for each building necessary to calculate the mutual assistance capability.

⑥ The estimation of resident information

What we can use now as residents' distribution information is the population census. However, the census currently made public is the aggregate for each city, ward, town, and village. Therefore, it is difficult to grasp the detailed population distribution such as the figure for each building.

Therefore, in this study, we used the micropopulation census prepared by Akiyama et al. [3], which is highly precise population distribution data, and prepared the household and residents' information (by age and sex) distributed to each building.

⑦ The estimation of mutual assistance capability (the number of people expected to be rescued)

From the survey of the Great Hanshin-Awaji Earthquake, we know that about 80% of the residents who were rescued from collapsed buildings were rescued by the residents living around the collapsed buildings [8]. Therefore, in our method, we supposed that the people living around a building rescue the residents of the building. Here, we define "around a building" as "within a 100 m distance by the road network." For the rescue capability of the residents, we consider the physical strength defined by age and sex

[5]National Land Numerical Information, Densely Inhabited District Data (http://nlftp.mlit. go.jp/ksj/gml/datalist/KsjTmplt-A16.html).

Table 4.2.2 The number of expected rescuers by age and gender

Age	Strength Male	Strength Female	Enforcement rate	Action rate Male	Action rate Female	Mean Male	Mean Female
10	1	0.85	0.228	0.76	0.24	0.1733	0.0465
20	1	0.76	0.228	0.76	0.24	0.1733	0.0416
30	0.96	0.76	0.229	0.72	0.28	0.1583	0.0487
40	0.93	0.73	0.298	0.72	0.28	0.1995	0.0609
50	0.9	0.72	0.228	0.72	0.37	0.1293	0.0607
60	0.84	0.7	0.191	0.74	0.26	0.1187	0.0348
70+	0.78	0.65	0.129	0.75	0.25	0.0755	0.0210

"10" means between 10 and 19 years old. "70+" means over 70 years old.

of each resident, and the distance from the building where the neighboring residents live. In other words, we supposed that the residents distributed in the buildings that are located within 100 m would come to rescue the people in the collapsed buildings. The rescue activity of the residents by age and sex, has been clearly shown in Table 4.2.2 from the survey results of the Great Hanshin-Awaji Earthquake [8,9]. By using the values of Table 4.2.2, we could estimate the mutual assistance capability of a building.

⑧ The estimation of fire-fighting capability (number of buildings where a fire was expected to be extinguished)

In this method, we estimate the fire-fighting capability, which is the number of buildings where a fire was expected to be extinguished, by the public fire-fighting capability (such as fire engines, and professional and volunteer fire-fighters) of the area where the building is located. First, by referring to the website of the Fire Chiefs' Association of Japan [10] and the National firefighting handbook, we found the distribution of fire-fighting facilities in Japan, the fire-fighting equipment and firemen placed there, and the number of the volunteer fire corps of the area. We then searched for the nearest fire-fighting facility by road network distance from each building, and calculated the distance from the fire-fighting facility, the number of fire engines, and the firemen and the volunteer fire corps that could be expected to come to each building. Using these figures, we could estimate the fire-fighting capability for each building.

By combining the results of the above estimates ⑦ and ⑧, we could estimate the fire-fighting capability and mutual assistance capability for each building in Japan.

Please note that the above calculation results are estimated figures, and there is a margin of error from the actual condition of each building.

According to Akiyama et al. [4], the aggregated results have been found to be very close to the actual conditions.

4.2.3 DAMAGE RISK EVALUATION FOR AN OCCURRENCE OF A LARGE-SCALE EARTHQUAKE BY MICRO GEO DATA

The estimated ①–⑧ figures for each building are shown in Fig. 4.2.3. Similar results are provided for approximately 60 million buildings in Japan. By combining these figures, we can estimate the damage risk after considering the damage condition of each building and the mutual assistance capability/fire-fighting capability (Fig. 4.2.4). However, the figures given to each building are estimated, so the results for each building are not necessarily exact. Therefore, the final damage risk results can be shown clearly by aggregating the results for each mesh, street, ward, etc.

In Figure 4.2.5, the damage risk results are shown for a seismic motion occurring on a winter evening, for which the probability exceeds 2% in a 50-year period. The areas where much damage is particularly expected are the south part of the Ishikari Plain, the Sendai Plain, the Echigo Plain, the Pacific coast of the Tokai region, and the Kochi Plain. In particular, along the Pacific coast of the Tokai region, areas where much damage is expected are distributed all along the coast. If a widespread disaster occurs, much more damage is expected under the current state of preparation.

Next, we focus on the three major metropolitan areas. Since earthquake proofing has proceeded pretty well in the Tokyo metropolitan area, there will be only a little damage by building collapse. However, very dense residential areas of wooden structures surround the Tokyo metropolitan area and these will be strongly affected by the spread of fires. Areas where more damage is expected are distributed in a donut-shape surrounding downtown Tokyo. On the other hand, damage in downtown Tokyo and along Tokyo Bay will be relatively small. A similar tendency is seen in the Osaka urban area. In the Nagoya urban area, such a donut-shape is not clearly seen, but in the western part which has wide rivers, we see the much damage. In these areas, it is estimated that the building collapse rate will be high, and the initial capability of responding is at lower. Therefore, urgent measures are required.

In this book, we have shown only the results when a seismic motion occurs with a probability exceeding 2% in a 50-year period. However, data users can also input any seismic motion to any area. Damage risk for a scenario based on input data of any seismic motion can estimated.

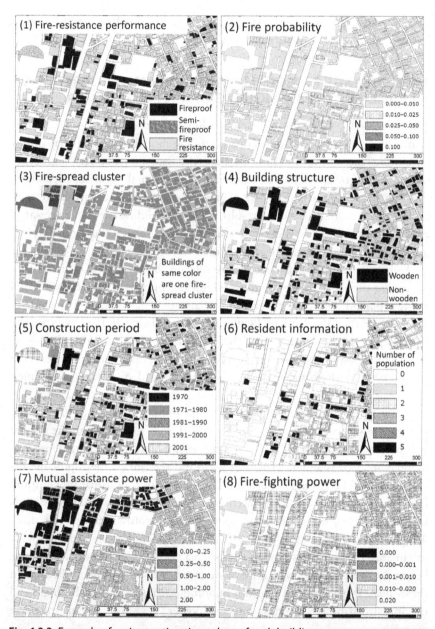

Fig. 4.2.3 Example of various estimation values of each building.

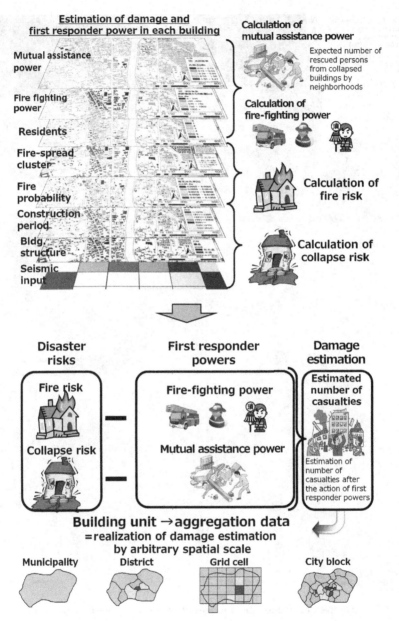

Fig. 4.2.4 Processing flow for earthquake damage estimation using various estimation values of each building.

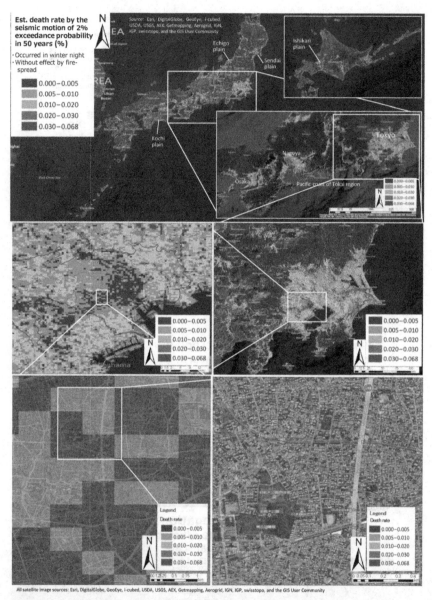

Fig. 4.2.5 Example of damage estimation results due to heavy earthquakes using micro geo data.

As shown above, the complex use of various micro geo data has enabled us to prepare basic data to evaluate the building collapse/fire-risk caused by large-scale earthquake damage and the initial capability of responding. Also, by using that data, a method to evaluate quantitatively the disaster risk and initial capability of responding for any input seismic motion and any aggregate unit was achieved. A comprehensive earthquake damage risk evaluation, considering damage risk and initial capability of responding, was also realized. As a result, the information such as (1) areas likely to be severely damaged, (2) whether the damage is by building collapse or fire, and (3) the initial capability of responding became available for evaluation of any part of Japan.

In the future, if these results were disclosed and shared properly with local governments, and also with community residents, our micro geo data could be utilized in a top-down manner like the regional disaster management planning by local governments, and also in a bottom-up manner such as the community residents' understanding of the danger and initial capability of responding of their area, sharing of damage risk, and voluntary preparation for a disaster. As a result, it is expected that it will lead to the area's resilience recovery/improvement with the local governments and residents united.

However, so far, data preparation for the basis only has been completed for all parts of Japan, and many issues remain. First, we need to be prepared for earthquakes that may occur at times other than winter evenings. For this, we need to have the data of detailed population distribution during the daytime. Since a tsunami disaster is also expected after a large-scale earthquake, it is desirable to make a scenario in which a tsunami is expected, by which we can estimate the damage risk and prepare an environment that will allow us to estimate the damage together with the building collapse/fire damage. Furthermore, it is necessary to discuss how to disclose and share the above results properly without misunderstanding by the local governments and residents.

As noted above, there are still some issues remaining. Still, the fact that the basic data for possible future nationwide earthquake damage and the initial capability of responding has been prepared would greatly contribute to the national disaster management policy, and further the planning of national strategies for building a resilient country.

REFERENCES

[1] Akiyama Y. Applications of micro geo data for urban monitoring. In: ICGIS2014 spatial big data technologies and applications for future society; 2014. p. 103–16.
[2] Akiyama Y, Sengoku H, Shibasaki R. Development of commercial accumulation statistics throughout Japan and utilization environment of them. Theory Appl GIS 2013;21(2):11–20 [abstract is in English].
[3] Akiyama Y, Takada T, Shibasaki R. Development of micropopulation census through disaggregation of national population census. In: CUPUM2013 conference papers, 110; 2013.

[4] Akiyama Y, Ogawa Y, Sengoku H, Shibasaki R, Kato T. Development of micro geo data for evaluation of disaster risk and readiness by large-scale earthquakes throughout Japan. Papers of research meeting on civil engineering planning, vol. 392; 2013. p. 1–19 [abstract is in English].

[5] Kato T, Sugata T, Akiyama Y, Sengoku H, Ogawa Y. Development of natural disaster risk evaluation platform for QALY. Proceedings of JAHES, vol. 16; 2013. p. 343–6 [in Japanese].

[6] Ogawa Y, Akiyama Y, Shibasaki R. The development of method to evaluate the damage of earthquake disaster considering community-based emergency response throughout Japan. GI4DM2013, TS03-1; 2013.

[7] Kato T, Hong C, Yalkun Y, Yamaguchi M, Natori A. A method for the integrated earthquake fire risk evaluation based on the single building fire probability applying to any different map-scale. J Soc Safety Sci 2006;8:279–88 [abstract is in English].

[8] Kawata Y. Prediction of loss of human lives due to catastrophic earthquake disaster. J Jpn Soc Nat Disaster Sci 1997;16(1):3–13 [abstract is in English].

[9] Tokyo Fire Department. Council for the 16th Tokyo fire prevention report; 2005 [in Japanese].

[10] Fire and Disaster Management Agency. The report concerning the development guideline of fire power; 2004 [in Japanese].

[11] Aichi Prefecture. Report of damage assessment in Aichi Prefecture; 2003 [in Japanese].

CHAPTER 4.3

Consensus Building for a Resilient Society: Utilization of Big Data

T. Kato
Institute of Industrial Science, The University of Tokyo, Tokyo, Japan

4.3.1 POTENTIAL OF USING GEO BIG DATA FOR IMPROVEMENT OF RESILIENCE

There are two major instances that suggest the potential of using geo big data for the development of cities and towns that are resilient to disasters.

The first instance is to use the data for community-based development of cities and towns that are resilient to disasters. In the event of a megahazard, self-support, community-based support, and governmental support implemented by all the people and organizations involved in prevention and mitigation of disasters are indispensable. Among these three forms of support, enhancement of community-based support is necessary, considering the limit of what we can expect from governmental support. Community-based support is mutual assistance or an effort at the community level, but the agent that provides community-based support—the local community—is not clearly defined. For this reason, although the importance of community-based support has been emphasized, how to establish such an effort in the community that will bring about practical, effective community-based support has been a social issue. Although there are some precedents, duplicating them across the nation requires development of a new approach that will produce something that encourages the initiation of community-based support and ensures a foundation to continue the effort. Geo big data has the potential to play that role.

The second instance is to use the data for measures taken by governments to prevent disasters. Anticipated earthquake damage is very commonly used as a basis of municipal disaster management plans. Anticipation of damage is indispensable for preparation of disaster-management plans and consideration of disaster-management actions, in that it presents pictures of our "enemy." A methodology of using engineering models has already taken root, but to make the plans and actions more appropriate, it is necessary to simulate damage in a more accurate, detailed manner. Conventionally,

Disaster Resilient Cities
http://dx.doi.org/10.1016/B978-0-12-809862-2.00014-0
135

data or statistics possessed by governments are used for anticipation. If geo big data is used, it may be possible to simulate disasters in a way that helps people to imagine in more detail how their communities could be struck by the disasters. Furthermore, as in the case of fire simulations, if the way damage develops differs, depending on the conditions given to the simulation, such as fire occurrence points or weather conditions, use of big data may make it possible to anticipate every situation that can potentially occur.

This chapter will introduce the possibilities and challenges of using geo big data that are examined from the two perspectives mentioned above, with reference to some instances.

4.3.2 GEO BIG DATA ASSISTING DEVELOPMENT OF CITIES AND TOWNS RESILIENT TO DISASTERS

4.3.2.1 "Self-Support, Community-Based Support, and Governmental Support" Indispensable to Society

These are the catchphrases that have been well recognized in Japanese society since the 1995 Great Hanshin Earthquake. They give an impression that society can be made more resilient by all those involved in disaster prevention, working together. The reality is that excuses, instead of efforts, have been made for governmental support, self-satisfaction has resulted from community-based support, and no policies have been prepared to encourage self-support. To take self-support as an example, the rate of fastening furniture is low, at a national average of 26.2% (Cabinet Office special opinion poll in Dec. 2009). Participants in disaster-prevention drills organized by neighborhood associations in communities are more or less fixed, which has remained an issue to address. On the other hand, 11 years after the 1995 Great Hanshin Earthquake, the rate of public school buildings renovated to be more earthquake-resistant was still just 51.8% (2006 survey of the Ministry of Education, Culture, Sports, Science and Technology). The school buildings are being renovated one by one owing to a limited public budget. Although the catchphrases, "self-support," "community-based support," and "governmental support" sound like good rhetoric, the flipside of these phrases is that the effort called for by each catchphrase depends on those called for by the other two and the catchphrases have kept society at a low level as a result.

We should aspire to realize self-support, community-based support, and governmental support that develop individually, autonomously, and continuously. Two necessary conditions are required to meet that goal: (1) that all of the agents involved in self-support, community-based support, and governmental support should understand possible scenarios of damage brought about by disasters that could occur in their community; and (2) that all of the agents should understand their own roles and what they are expected to achieve. When these two necessary conditions are satisfied, the agents of community-based support understand not only a picture of their "enemy," but also the limitation of governmental support, as well as the limitation of self-support and governmental support combined. They will also learn to understand, without difficulty, issues that cannot be solved by any of the three efforts, or issues that need to be addressed through community-based support. It is expected that community-based support will have the qualities of endogenous and autonomous development, based on the basic recognition described above. This is the way self-support, community-based support, and governmental support should be, and what we should aspire to, and this should be well recognized by Japanese society [1] (Fig. 4.3.1).

4.3.2.2 Roles of Geo Big Data, Past Instances, and Future Possibilities of Using the Data

The following instances are precedents of using geo big data in community-based development for disaster prevention. In each of the instances, geo big data functions as a technology assisting to implement "self-support, community-based support, and governmental support" in the expected manner described earlier, and the data works as a tool that assists in one of the necessary

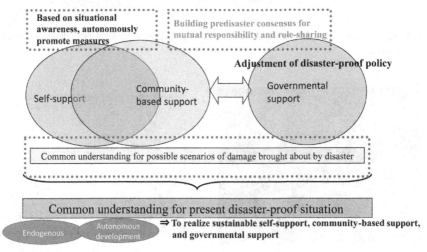

Fig. 4.3.1 Ideal form of "self-support," "community-based support," and "governmental support."

conditions—understanding the way a community could be subjected to damage. The tools referred to in Figs. 4.3.2–4.3.4, for example, are typical tools.

① System that combines a simulation technology and geographic information system (GIS) to assist community development (Figs. 4.3.2 and 4.3.3):
  ①-A planning support system for disaster-proof communities [2],
  ①-B support system for understanding flood hazard [3];

② Tool that uses Google Earth to illustrate the hazards and risks in a particular area (Fig. 4.3.4);

③ Tool that assists in understanding hazards to which local communities are exposed, with the help of augmented reality (AR) technology (Fig. 4.3.5; Ref. [4]).

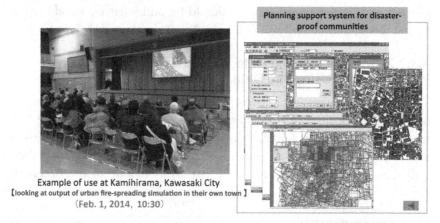

Planning support system for disaster-proof communities

Example of use at Kamihirama, Kawasaki City
【looking at output of urban fire-spreading simulation in their own town 】
(Feb. 1, 2014, 10:30)

Fig. 4.3.2 Utilization example of the planning support system for disaster-proof communities.

Fig. 4.3.3 Utilization example of the support system for understanding flood hazard by residents.

Fig. 4.3.4 Visualization of flood hazards and risks simulation results in a particular area by general display tool (Google Earth).

All of them have been used practically to develop communities for disaster prevention, not only to clarify potential risks to which communities are exposed, but also to bring about a useful effect on the way the development of communities is put into practice.

Support system ①, based on the combination of the simulation technologies, is useful for understanding chronologically how damage develops in the event of a fire or an extensive flood, for example. By inputting any fire occurrence point, any weather condition, or any part of a levee to be broken as simulation conditions, we will understand how fire will spread or how land or buildings will be flooded. The system helps us to understand how differently damage will develop under different conditions and understand it more deeply than risk maps or hazard maps. Furthermore, the system ①-A, which assists community development for disaster prevention, can simulate how redevelopment of buildings and roads mitigates

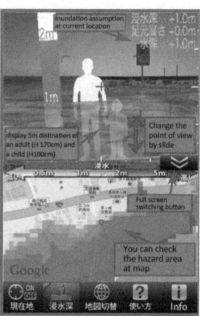

Fig. 4.3.5 Example of the display system for understanding hazards to which local communities are exposed, with the help of AR technology.

damage. Fig. 4.3.5 shows a scene of how the system ①-B is used practically. It is common that someone with expertise operates the system to explain disaster risks to residents, but this is a pioneering instance in which some residents who played an active role in community-based support operated the system themselves and explained the risks to their neighbors. Leaders of neighborhood associations, who had participated in a training course and learned how to operate the system beforehand, organized a workshop for community development for disaster prevention. This instance is thought to suggest the following two future possibilities.

First, the instance demonstrates that ordinary residents, if they are not experts but are highly motivated, can learn how to use highly sophisticated technology. The leaders of neighborhood associations are not experts with high-level IT skills, but ordinary elderly people. They demonstrated that they could learn the necessary skills if they were willing to do so. Second, it was found that the explanation by residents, rather than by experts, resulted in better understanding. The data used in this instance is government data of urban areas that contain attributes of the areas. However, at most, the information about communities that external experts can interpret from the attributes of such an area is the structure, application, and number of stories of the buildings there. In contrast, each of the communities has detailed data unique to it that cannot be expressed in the form of digital data. For example, let us suppose that there is a piece of private personal information that there is a household in a building in which two elderly people live alone and one of them fell ill last month and has been in bed in this room since then; the community has such private information, in the natural manner that the people there "know it." By considering private information that was embedded in a community and would never be released, and by adding it to the explanation of natural disasters, the residents in the community better understood the challenges of their community trying to prevent disasters. This suggests that the development of highly sophisticated technology by experts and provision of the technology to communities alone are insufficient, and that such technology may be of use to communities only if it is combined with another technology to be adapted to each of the communities. The instance suggests a way of using geo big data in the future. It may look like pioneering, and is extraordinary in a sense, but it can be regarded as an ideal model that can be introduced to other communities.

System ② uses Google Earth, a general-purpose tool, to visualize static map information that concerns hazards and risks, and the use of the general-purpose tool has been advantageous in various ways in community development for disaster prevention. A direct effect of using this system is that people can imagine vividly and understand risks of natural disasters by looking at aerial imagery and street view which comes with Google Earth. An indirect effect is that there has been a change in communication between residents, and that the introduction of the tool has brought in newcomers. In this case, when Google Earth was introduced, junior high school students were asked to take part as new participants. When junior high school students participate in meetings organized mainly by elderly people, the roles of the participants tend to be fixed, with elderly people teaching students. However, in this case, the strengths of the two groups—junior high school students who skilled in using the tool but who do not have knowledge, and elderly people who have knowledge but do not have the skill to use the tool—were combined. The participants from the different groups developed discussions on equal terms. Another aspect to note is that the workshop organizer informed in advance that the tool would be used and this announcement elicited a response from the neighborhood association, which encouraged and persuaded people with IT skills to participate in the workshop. This suggests that the introduction of new tools or technologies can bring in those who have not taken part in the workshop before.

The junior high school students and people with IT skills who took part in the workshop continue to play an active role, as new participants, in the community's disaster-prevention activities.

System ③ is a much more accessible application, which is equipped with AR technology to help people to see disaster hazard risks in the form of map information that is superimposed over images from smartphone or tablet cameras, and presents hazard risks of a certain place in a more straightforward manner. The system, the purpose of which is to help people to understand disaster hazards and risks more graphically, has been in use in the workshops referred to above. In addition, secondary effects are hoped for. This system has been released to the public to be downloaded by anyone. Drawing attention of more people through increased accessibility to disaster risk information is considered another secondary effect of using the system.

For the above reasons, the use of data on such a small scale that visualizes individual buildings (ie, micro geo data) helps people to understand disaster hazards and risks better, sets off a new action as part of community-based support, and contributes to the nurturing of sustainable community-based support. The introduction and application of this system to other areas are also expected.

What should be noted about the systems presented here is that all of the systems use the micro geo data managed by local governments, which continues to be collected through additional surveys in cooperation with these governments. It should also be noted that these systems can be introduced only to local governments that have usable micro geo data. To introduce and use the systems nationwide, it is essential to prepare micro geo data in every local government and make the data accessible to anyone anytime. Building a natural disaster risk information platform is expected to contribute to making the data available.

4.3.3 GEO BIG DATA RAISING ACCURACY OF DISASTER-PREVENTION PLANS

4.3.3.1 What Does Anticipation of Earthquake Damage Mean?

Anticipation of earthquake damage, through which we learn about our "enemy," serves as a basis for preparing municipal disaster management plans. Efforts have been made to innovate the technology for anticipation through studies and research done by government, and the technology can be said to have already reached a certain level. To put it simply, anticipation of earthquake damage collects data on urban areas, builds a model of a city in the computer, chooses the location of the source of an earthquake, uses an engineering model to calculate how much damage will be done and how extensively the damage will spread in the city, and simulates how the city will be damaged. The extensively collected data includes the data of ground, underground infrastructure and constructions like buildings and bridges, and social data like population. Objects from which the data is collected are not only those managed by governments but also others operated by utility companies and the like.

For example, the Tokyo Metropolitan Government published a report on anticipated earthquake damage in 2012. Four earthquakes which occur in different locations were simulated to find various aspects of damage, such as shaking of the ground, soil liquefaction, damage to constructions, fire damage, number of evacuees, and number of people unable to return home.

4.3.3.2 Role of Micro Geo Data Expected to Contribute to Anticipation of Earthquake Damage

To enhance the quality of disaster-prevention plans, it is necessary to improve the accuracy of damage anticipation, which depends on the accuracy of engineering models for computation and of data. Obviously, to anticipate damage more correctly, engineering models and data need to be made more precise. Data collected in the form of mesh data or aggregated on a scale of individual districts (a group of city blocks) of a municipality are often used. There is no example of anticipating damage on the scale of individual buildings. Governments possess data of individual buildings, in the form of fixed asset tax records, for example. However, these are protected against being used for unintended purposes, and the government is prevented from divulging personal information. Therefore, using the data in its present form is restricted. An assessment of earthquake damage risks recently carried out by the city of Nagoya used data aggregated on a scale of individual city blocks, each of which is far smaller than a district. Supposing that there are limitations to raising the accuracy of engineering models, improvement of data accuracy is indispensable to increasing the accuracy of anticipation of earthquake damage. As explained earlier, to put self-support, community-based support, and governmental support into practice, it is necessary to learn about our enemy. To understand with a sense of reality how the enemy will behave, it is expected, therefore, that data at the level of individual buildings will be available.

4.3.3.3 Big Data Contributes to Raising Accuracy in Anticipation of Earthquake Damage

The process of anticipating earthquake damage involves a number of engineering models, whose computational accuracy may vary among models. There are not only error in engineering models, but also uncertainty in phenomena. Anticipation of the death toll caused by fires following an earthquake depends on where the fires break out. The incidence of fires can be quantitatively evaluated, but it is essentially impossible to locate where fires will break out in the event of an earthquake.

Anticipation of the death toll greatly depends on where fires break out. For example, let us think about one of the worst situations, in which fires occur in a manner that encircles many people. The fires will spread over time and the area burned by the fires will make a donut shape, and every part of the urban area inside the fires will eventually be burned up. It is easy to suppose that the anticipated death toll will be tremendously large. On the other hand, if the density of the fire occurrence points is the same but fires develop in a manner that is lucky for evacuees, the anticipated death toll will not be so large because the fires develop approximately one-tenth as fast as people walk. The Tokyo Metropolitan Government anticipated earthquake damage and calculated that as many as 4100 people would be killed in fires, but this figure should be interpreted as a benchmark and the actual figure in the event of an earthquake could differ substantially from that.

Anticipation of damage these days often does not use simulation but past instances for an estimation through a simple technique. Considering the conventional environment, this has to be accepted in order to save cost, data, and computer resources and data storage resources. However, the age of big data has arrived, with detailed data, plenty of computer resources, and data storage resources easily available. If we take advantage of this environment, it is possible to increase the accuracy in anticipation of damage.

This paragraph will introduce an experimental study to discuss its future potential. Fig. 4.3.6 is a sample of output from a fire simulation system that uses data of individual buildings, while Fig. 4.3.7 is an example of output from the same system to which an evacuation simulation is additionally input. The figures show the result of simulations of evacuations of one million people in total. As much as 24 GB data are created per simulation. In a current big data environment, simulations can be repeated with different patterns of fire occurrence points. Fig. 4.3.8 is a result of computation that produced 3000 patterns of fire occurrence point distribution, on the condition that the number of fire occurrence points is fixed and fires occur

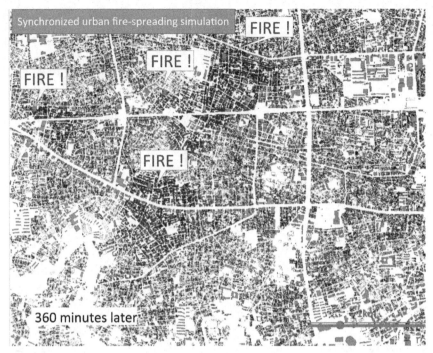

Fig. 4.3.6 Output from the simulation system of postearthquake urban fire spreading (Suginami and Nakano, Tokyo).

randomly in relation to the incidence of fire in individual buildings, and presents the frequency distribution regarding the number of people who passed areas near fires caused by an earthquake, exposed to heat. This number of people is thought to correspond to those who get lost and are killed while trying to escape the fires in the event of the earthquake. In this evacuation model, the number amounted approximately to 400–500 on average, with the distribution spreading out largely on the right. Depending on distribution patterns of fire occurrence points, the number of the evacuees who passed dangerous areas can be between 4000 and 7000, which is 10 times more than the average.

Previous attempts at anticipating damage gave figures more or less close to the average—that is, a benchmark value. However, taking advantage of a big data environment revealed the possibility that many evacuees may be killed, depending on conditions. The existing techniques for anticipating earthquake damage, by their nature, cannot point out such a possibility. We have to wait for results of future research to evaluate quantitatively the possibility of occurrence and analyze the structure of risk factors. However, it

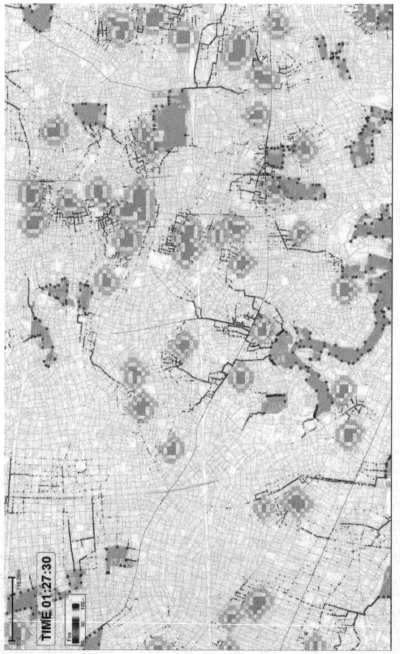

Fig. 4.3.7 Output from combined simulation of urban fire spreading and evacuation (Suginami and Nakano, Tokyo).

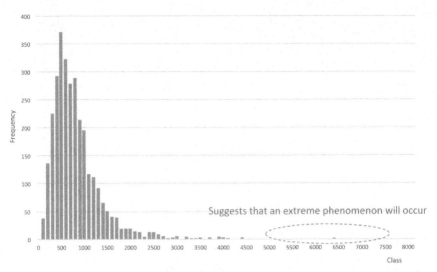

Fig. 4.3.8 The frequency distribution of causality on the way to evacuation area by urban fire spreading.

is not an exaggeration to say that a big data environment will increase the accuracy of anticipating earthquake damage and surely contribute to realizing more appropriate municipal disaster management plans.

REFERENCES

[1] Kato T. Basis of community-based disaster preparedness and mitigation and perspective for disaster-proof communities. Urban Housing Science, vol. 83, p. 60, Tokyo. 2013 [in Japanese].

[2] Promotion and administration committee of the planning support system for disaster-proof communities. Disaster mitigation community development planning support system. http://www.bousai-pss.jp/ [accessed 01.12.14].

[3] NPO Oh! Safety & Amenity Machizukuri. Report on community-based activities for disaster preparedness with flood inundation simulation, study of urban renaissance supported by the government. https://www.kantei.go.jp/jp/singi/tiiki/toshisaisei/05suisin/kantoh/04suisin/h16/33.html [accessed 02.25.15].

[4] Fire and Disaster Management Agency. Annual report on research and development for fire and disaster prevention. http://www.fdma.go.jp/neuter/topics/fieldList4_2/pdf/seika_jirei05-h2703.pdf [accessed 01.12.14].

CHAPTER 4.4

Resilient Regional Grand Design Based on Quality of Life

Y. Hayashi, H. Kato
Graduate School of Environmental Studies, Nagoya University, Nagoya, Japan

4.4.1 INTRODUCTION

The occurrence of the Great East Japan Earthquake and climate disasters, including torrential downpours, is raising awareness about disaster prevention and mitigation among Japanese people. Due to the raised awareness, there is concern that measures against disasters have not been carefully examined in discussions of national-level land planning or urban planning. However, because Japan has been frequently subjected to damage from earthquakes and extreme weather, the implementation of such measures is long overdue, so a detailed method to implement them should be introduced soon. In addition, climate change due to greenhouse gas (GHG) emission is anticipated, and there is concern that typhoons, sudden downpours, tornadoes, etc. could be more frequent and intense. Furthermore, since seismic and volcanic activities are expected to increase, it will be necessary to respond to enormous disasters brought about by these phenomena.

On the other hand, the aging of society is rapidly developing in Japan, which is facing great difficulties in formulating an appropriate reaction. When there are very limited usable resources, an important reaction in the first instance to take against growing natural disaster risks and increased vulnerability of our social system is to prevent urban and residential areas from expanding further.

To this end, it is necessary to establish a method to choose the urban and residential places to maintain and renew to collect other urban and residential areas that result in an oversupply, as well as a method to determine a detailed policy on how to choose urban and residential areas to collect the population and functions of unchosen areas and decide how to encourage people to retreat from there. A fundamental premise to establish the method is to stop the system of designing and developing buildings and infrastructures independently of each other, which has been adopted by Japan, and discuss and consider comprehensive measures, including the method of locating buildings. This chapter will give a general account of a methodology, being developed by the author and others, to obtain spatial information that will be taken into account for the discussion and examination.

4.4.2 RESILIENCE AND SUSTAINABILITY QUANTITATIVELY EXPRESSED BY TRIPLE BOTTOM LINE

This book discusses how a resilient nation should be designed. To include that discussion in preparation of actual land-use planning, it is necessary to evaluate quantitatively how land use affects resilience. However, it is insufficient to discuss the design of the nation from the perspective of resilience alone. "Sustainability" is considered a more essential and comprehensive concept, as another perspective that should be included in the discussion.

The idea of sustainability became widely known by the public after the World Commission on Environment and Development released a report called "Our Common Future" in 1987 and used the term "sustainable development" in the report: "Humanity has the ability to make development sustainable to ensure that it meets the needs of the present without compromising the ability of future generations to meet their own needs." Later, the eight Millennium Development Goals (MDGs), prepared based on the Millennium Declaration adopted by the United Nations Millennium Summit in 2000, were set as eight goals, including that of ensuring environmental sustainability, to be achieved by 2015. The United Nations Conference on Sustainable Development held in 2012 (Rio+20) shared a common recognition that it was important to build a society based on three parts—economic, social, and environmental (called the triple bottom line, or TBL)—and agreed to start a process to prepare, as well as take action to achieve, the Sustainable Development Goals (SDGs) as new follow-up goals to the MDGs. The SDGs consist of 17 goals and 169 specific targets, and many of them call for consideration of resilience, bearing in mind the need to adapt to the rising temperature and sea level and the extreme weather events brought about by climate change. This means that resilience is widely recognized as one of the elements that heighten sustainability.

Based on these discussions, this book will call a possibility to continue for more than decades "sustainability" and an ability to deal with damage caused by a huge disaster that could happen once in more than several decades "resilience," and quantitatively expresses the two concepts in the form of changes in the TBL of the human community, which consists of economy, society, and environment. In this chapter, economy, society, and environment are simply defined as the cost of maintaining urban areas, the level of quality of life (QOL), and emissions of GHGs, respectively.

The condition desirable for the TBL is that a high-level QOL can be obtained at a reduced cost and with a smaller amount of GHG emissions. Sustainability can be interpreted to mean that the TBL moves toward this condition in the long term. On the other hand, in the event of a huge disaster, the level of QOL will drop temporarily but sharply, and a large amount of cost and GHG emissions will be paid and produced respectively as a result. Resilience can be defined as the ability to contain the amount of the paid cost and produced emissions at a low level. If we follow the above train of thought, once we succeed in measuring and anticipating the condition toward which the TBL will move in the future and the amount of its changes caused by huge disasters, it will be possible to apply the measurement and anticipation for the consideration of how to design a resilient and sustainable nation.

4.4.3 CHANGES IN QOL CAUSED BY HUGE DISASTER, AND RESILIENCE

This book discusses the level of QOL, which represents social aspects particularly among the elements of the TBL, and examines the quality of resilience against huge disasters. Immediately after a huge disaster, while priority is given primarily to saving human lives and making arrangements so that evacuees can continue their lives, the priority of paying attention to cost and burden on the environment is low. So it is not necessarily inappropriate to discuss QOL alone, but it should also be noted that all elements of the TBL should be considered when a society at peace takes precautionary actions and implements programs against disasters or when reconstruction begins or is under way.

Fig. 4.4.1 is a schematic diagram, which shows QOL decreased by huge disasters. The figure gives QOL in the form of life year indexes [1]. In the assessment of nations and societies, QOL is often represented quantitatively by economic values, but in the field of medicine, to which the concept of QOL was introduced earlier than to other fields, it is commonplace to express the drop of QOL attributable to illness and injury in the form of the disability-adjusted life year (DALY). In the same manner, by using the quality-adjusted life year (QALY), which is a measure of the extent

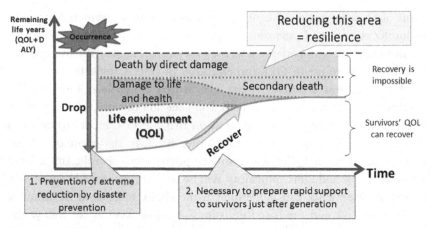

Fig. 4.4.1 QOL decrease by huge disasters.

to which the decrease in the standard of living is perceived to shorten life years, the sum of the QALY and the DALY can be treated as QOL in which death, injury, and decrease of the standard of living are taken into account.

Chapter 3 gives an account of the dropped QOL of those who underwent the Great East Japan Earthquake, but the drop of QOL has not been quantified yet, using the QALY. The calculation requires clarification of how greatly the elements of QOL will influence the QALY, but the extent of their influences depends on individual persons' senses of value, which vary among individuals. However, people's senses of value are considered to be similar to some extent if the attributes of the people are similar. For example, ages have a big impact on QOL. If different senses of value between different generations can be clarified through questionnaires, etc., it will be possible to take account of the impact of an aging society to analyze how the QOL index will change.

Fig. 4.4.2 illustrates the way QOL drops sharply from the normal-time level when a huge earthquake happens, and the way it begins to recover slowly afterward. However, because death strips casualties of all their life years, their QOL continues dropping until the time their life years are supposed to come to an end. Injury and dropped standard of living recover step by step, but if a person suffers a sequela or their life does not fully recover, QOL also continues declining for the rest of their life, as in the case of their death.

Here, resilience can be defined as "smallness" of the total amount (integration value) by which QOL is decreased by a disaster, and it can be enhanced if its decline in the event of the disaster is inhibited and it recovers quickly.

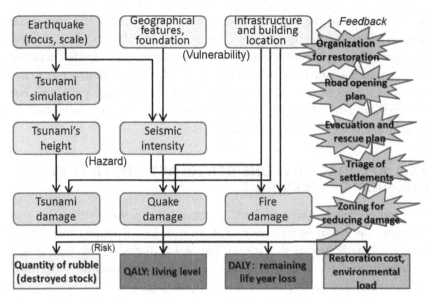

Fig. 4.4.2 Flowchart for qualifying social resilience against a big earthquake.

The life-expectancy value (total amount of QOL decline) evaluated in such a manner essentially affects the assessment value of long-term sustainability. It means that, if the incidence (frequency) of an anticipated huge disaster and total amount of QOL increase can be anticipated and if residents understand them properly, the risk assessment value of that huge disaster expressed as a product of the incidence multiplied by the total amount of the increase can be regarded as a factor that reduces the normal-time level of the QOL index. In practice, it is thought that such a situation can hardly occur because of difficulty in anticipation and understanding, but if the accuracy of anticipation is improved, and if efforts are made to disseminate the result of anticipation and residents understand it, people will be encouraged to choose places of residence, taking account of the places' resilience to huge disasters, and activities preparatory to disasters will be promoted.

It should be noted that, when considering resilience to huge disasters, it is important to take account of the extent to which QOL drops when such a disaster occurs, in addition to the total amount by which QOL drops. If QOL decreases by an extremely large amount, a substantial amount of time and money are required to rebuild the damaged community, or the condition of the community could be so catastrophic that the community cannot be reconstructed any more. If it is predicted that such a situation could occur at a high probability, resilience is considered extremely low and measures to avoid the situation should be examined. Disasters that could happen extremely infrequently, or once in a thousand years like the Great East Japan Earthquake, may have a small assessment value if it is calculated on an annual average basis, but once it occurs, it will be a destructive strike; therefore, it may be necessary to take measures against even such large but infrequent disasters. How to address such infrequent mega risks is also important when we consider resilience.

4.4.4 APPROACH TO CONSIDERING ACTIONS TO PREVENT QOL FROM DECREASING

The authors are working to build a spatial information system that can compute the extent of resilience of all areas in Japan, divided into a level of as small as 500 m mesh patterns. Fig. 4.4.2 shows a flowchart, which explains an example of how this system can be used to examine what measures can be taken to improve resilience to earthquakes.

First, the simulation takes account of the landforms and ground conditions based on the source zone and scale of an anticipated earthquake, and then it estimates the extent of shaking and height of a tsunami caused by the earthquake. This information is superimposed on the information about the distribution of buildings and population so that the damage caused by shaking, soil liquefaction, fires, and tsunamis will be estimated as the casualty toll and seriousness of damage to buildings. The damage in such forms will be converted to how much the DALY is increased as the casualty toll increases and how much the QALY is decreased as the standard of living drops, so that how much QOL drops will be calculated. The amount of the increase/decline varies depending on changing postdisaster conditions, such as how well evacuation and rescue are implemented, how quickly infrastructure and buildings are rebuilt, and how fast roads are opened; monitoring how the amount will change over time enables us to understand the level of resilience. It is also possible, at the same time, to estimate the amount of debris produced, cost of recovery, and burden on the environment, when infrastructures and buildings are destroyed.

This system can be applied to other disasters, including flooding, as well. The result of estimation makes it possible to consider various measures taken to raise resilience. The following actions or programs are specifically anticipated: disaster-reduction zoning that restricts land use in areas expected to be damaged severely; determination of order in which to open roads, by identifying which roads, if damaged, will make evacuation, rescue, and traffic impossible, and clarifying which section of the roads will particularly obstruct the movements of people/goods for all actions; performance of local-community triage to identify which districts will be damaged too seriously for the residents there to find an evacuation place within the districts, in order to evacuate the residents quickly outside the districts; and planning of community reconstruction.

REFERENCE

[1] Kachi N, Kato H, Hayashi Y. A computable model for optimizing residential relocation based on quality of life and social cost in built-up areas. J East Asia Soc Transp Stud 2007;7:1460–74.

CHAPTER 4.5

Paradigm Change in Flood Protection Strategies for Enhancing Resilience

N. Kachi
Graduate School of Engineering, Kyushu University, Fukuoka, Japan

Most land in the Netherlands is in flat, low-lying reclaimed areas and is vulnerable to flooding. In 1953, large storm waves occurred in the North Sea, claimed the lives of about 1800 people and destroyed the houses of about 200,000 residents. It was the largest flood damage in the history of the Netherlands. Additionally, the prediction that the sea level will rise and that flooding will increase owing to climate change has posed a danger in recent years that the country will be more vulnerable to flooding.

This chapter will give general information about Room for the River—a Dutch government flood protection policy closely related to land management and land use. The policy was adopted to maintain the safety level of land from flood damage that was expected to increase due to climate change. It will also present not only structural measures, but also land-management and land-use policies taken by the Netherlands to show how the country tries to enhance the safety level of land from flooding. In addition, the chapter will examine how Japan has developed its policies on land management, land use, and flood control, and what the country will do for flood control in the future in response to climate change, in comparison to Room for the River.

4.5.1 ROOM FOR THE RIVER PROGRAM IN THE NETHERLANDS—MORE SPACE FOR WATER

In the Netherlands, ring levees were set up in extensive areas from the 13th century to the 14th century and continued to be strengthened and heightened. Therefore, Dutch people took it for granted that reinforcing levees would protect the people from floods, and continued promoting the development of land through the reinforcement of levees until the late 20th century. The people never recognized that levees that continued to be heightened would do tremendous damage and make the society more vulnerable when they were broken.

However, as climate change issues surfaced in the 2000s, the increased risk of water disasters attributable to climate change began to be taken more seriously in the Netherlands. In 2006, the Royal Netherlands Meteorological Institute (KNMI, or Koninklijk Nederlands Meteorologisch Instituut) released the KNMI Climate Change Scenarios 2006 for the Netherlands, which included predictions that the sea level would rise by a maximum of 85 cm above the 1990 level by 2100 and the flow rate of flood at the Rhine gauge at Lobith—a small village on the banks of the Rhine—would be increased.

To prepare for climate change and increased risk of flooding, the Dutch government prepared the National Spatial Strategy in 2006, which gave guidelines for well-controlled development of flood plains, and the Spatial Planning Key Decision "Room for the River (SPKD)." The Room for the River Program, designed to promote river maintenance programs and reduce the vulnerability to flooding, was approved in 2006.

Simply speaking, the Room for the River Program is aimed at making adjustments to accommodate the flood flow volume, which is expected to be increased by climate change, not by making levees taller but by enlarging the river width, utilizing river beds, providing flood channels, expanding retarding basins, building water storage facilities, etc. Retarding basins are expanded as agricultural land that was previously protected by same safety level of housing and industrial land. In return, these retarding basins will be compensated when flood damages happen. The policy is a landmark change for the Netherlands, which had built and maintained levees for a long time to expand its useable land area.

Fig. 4.5.1 shows changes in the way the land surrounding the city of Arnhem has been used, and shows that urbanization has advanced from the 1800s to 2000s thanks to augmentation of flood control.

Room for the River requests a change of protection policy of such expanded urban areas that previously depended on only strengthening conventional flood control measures. It also calls for the expansion of river spaces to accommodate more floodwater, the expansion of river spaces to be used as retarding basins that reduce the amount of overflow, and the limitation and relocation of residential areas to prepare for increasing external force of flooding (Fig. 4.5.2).

The impact of the rising sea level on seawalls is also being discussed in the Netherlands. The storm waves that occurred in the North Sea in 1953

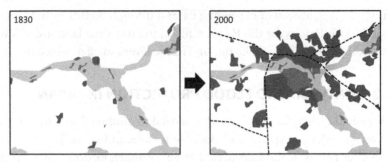

Fig. 4.5.1 Urbanization around Arnhem in the Netherlands. *(Data from Ministry of Transport, Public Works and Water Management, the Netherlands (Ministerie van Verkeer en Waterstaat, V&W). Spatial Planning Key Decision "Room for the River", investing in the safety and vitality of the Dutch River basin region; 2011).*

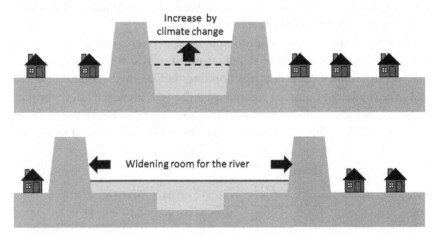

Fig. 4.5.2 Creating more space for the river corresponding to climate change.

caused tremendous damage to the country, with the lives of 1836 people claimed and an area of approx. 2000 km² (approx. 1/20 of the country's land) submerged [1]. As a result, the seawalls on the North Sea are designed to secure safety against a severe flood that could occur once in 10,000 years. The safety is predicted to be decreased by factors attributable to climatic change, including the rise of the sea level. It was decided that beach nourishment would be implemented by 2050 to prevent flooding along the North Sea coast.

Historically, the Dutch coastal areas were reclaimed and seawalls were set up in front of the reclaimed areas, but the country has chosen to use the areas subjected to beach nourishment in the process of reclamation to reduce the energy of waves, instead of expanding its land through reclamation. This corresponds with Room for the River, which intends to use land inside levees for retarding basins (ie, to make the land inside levees smaller) to secure safety.

4.5.2 LAND USE AND FLOOD PROTECTION IN JAPAN

As described earlier, the Netherlands has tried to confine floods to the space of rivers in order to expand land available for use, and so has Japan.

To take Tokyo in the Edo period as an example, its downtown around Edo Castle (currently the Imperial Palace) was protected by letting floods occur in the upper course of rivers flowing through the city. Chujo Levee, once set up in the area of the current Kumagaya City, worked for such a purpose. The lower course of the rivers also had Nihon Levee and Sumida Levee, and the areas that are the current wards of Kita, Arakawa, Katsushika, and Adachi played the role of retarding basins.

A sharp growth of population and rapid modernization of economy after the Meiji period called for an expansion of Tokyo's urban areas. With the construction of the Ara River floodway before the Second World War and of seawalls and pumping stations after the war, increased safety through flood control was provided to the middle course of Ara River and the eastern part of Tokyo, which advanced urbanization of these areas (Fig. 4.5.3).

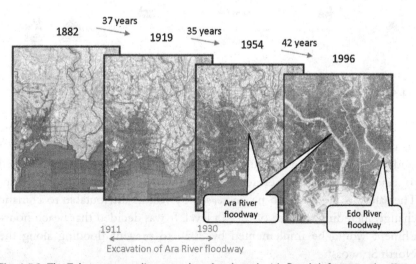

Fig. 4.5.3 The Tokyo metropolitan area has developed with flood defense works. *(Data from National Institute for Land and Infrastructure Management (NLIM). The utility of infrastructure in the Tokyo Metropolitan Area. Technical note of National Institute for Land and Infrastructure Management, no. 293; 2005. ISSN: 1346-7328. p. 4 [in Japanese]).*

The advance of urbanization in other large cities took a similar course; while flood control measures were taken, more people and assets were concentrated in cities, and efforts were made to confine floods to the space of rivers to protect people and assets. It was believed that people living in cities obtained safety and a good standard of living, but did they? Fig. 4.5.4 shows changes since 1950 in the amount of money spent on flood control, area of flooded places, and amount of damage caused by floods. It is obvious that flood control measures that have been taken have surely reduced the flood-inundated areas. On the other hand, the economic damage caused by floods does not show any downward tendency in spite of a decrease in area of the flood-inundated areas. The density of assets in places exposed to the risk of flooding has increased, and once such places are flooded, serious financial damage will be done, and it can be said that such is our society.

In such a situation of our society, it is pointed out that flooding, sediment disasters, storm waves, etc. may occur more frequently and become harsher in response to such phenomena as the rise of sea level, increased frequency of downpours, and intensified violence of typhoons. In response, the Minister of Land, Infrastructure, Transport and Tourism (MLIT) requested that the Panel on Infrastructure Development in the ministry discuss these issues, and the panel prepared a report, "Climate change adaptation strategies to cope with water-related disasters due to global warming," in 2008.

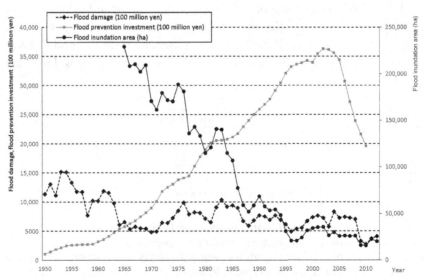

Fig. 4.5.4 Change in flood inundation area, flood damage, and flood-prevention investment (five-year moving average). *(Data from Statistics of flood damage (MLIT)).*

The report [2] made a proposal for future flood-control actions, according to the following basic concept: "a sustainable 'water-disaster adaptation society' should be built through an appropriate combination of adaptation and mitigation, while revising the present social structure and aiming to build a society coexisting with nature and realizing high energy efficiency in addition to safety and security." Its proposal said: "therefore, future flood control policies should be multilayered. In addition to traditional 'flood control policies to secure safety at the river level' through river improvement and the construction of flood control structures to meet target discharge levels for past and current river projects, 'flood control policies to secure safety at the basin level' through preparation for possible increase in excess floods should also be implemented. Multilayered flood control policies will be effective to flexibly cope with possible floods of different scales in each basin. Basin-based flood control measures should be proactively promoted, while comprehensive flood control measures which have been implemented should be further enhanced." Fig. 4.5.5 illustrates the ideas in the proposal.

The report urges a change of the flood-control and land-management policy, which was narrowly focused on expansion of areas protected against floods (ie, areas for residences), and encourages a review of the conventional system of land use to ensure resilience of such residential areas to be protected. This concept coincides with the philosophy of the Netherland's Room for the River.

Fig. 4.5.6 is an instance of securing retarding basins and securing flood safety level of communities by collecting scattered communities on the Kokai River, a tributary of the Tone River, after an extensive flood in 1986, to enhance safety both in this area as well as downstream areas. There are

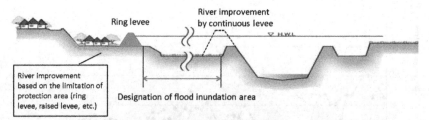

Fig. 4.5.5 Promotion of flood control in step with land use regulation and guidance: Land use and regional development should be promoted on the assumption that a flood will occur to cope with floods exceeding the design levels of structures. *(Data from Panel on Infrastructure Development, Ministry of Land, Infrastructure, Transport and Tourism (MLIT). Reference of "climate change adaptation strategies to cope with water-related disasters due to global warming (policy report)"; 2008. p. 58).*

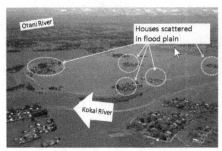 

Hakojima area damaged by the flood in 1986   Hakojima retarding basin constructed in 1990

Fig. 4.5.6 A new concept for urban development: compact community easier to implement flood control measures. *(Data from Panel on Infrastructure Development, Ministry of Land, Infrastructure, Transport and Tourism (MLIT). Reference of "climate change adaptation strategies to cope with water-related disasters due to global warming (policy report)"; 2008. p. 58).*

other similar examples in Japan, and it is hoped, from the perspective of Smart Shrink, that this approach will be applied to more instances in the future.

As discussed earlier, it is considered not coincidental but inevitable that both the Netherlands and Japan are calling for a paradigm change, from a conventional expansion-oriented manner of land use to the Smart-Shrink manner of land use based on selection and concentration, in order to make their land more resilient to the growing external force of natural disasters.

REFERENCES

[1] Deltawerken online. The flood of 1953, rescue and consequences. http://www.deltawerken.com/Rescue-and-consequences/309.html. [accessed 02.26.2016].
[2] Panel on Infrastructure Development, Ministry of Land, Infrastructure, Transport and Tourism (MLIT). Climate change adaptation strategies to cope with water-related disasters due to global warming (policy report); 2008. http://www.mlit.go.jp/river/basic_info/jigyo_keikaku/gaiyou/kikouhendou/pdf/draftpolicyreport.pdf. [accessed 02.26.2016].

CHAPTER 4.6

Smart Shrink Strategy and the Fiscal System

K. Tsukahara, N. Kachi

Graduate School of Engineering, Kyushu University, Fukuoka, Japan

4.6.1 INCREASING VULNERABILITY IN VIEW OF THE SOCIO-ECONOMIC CONDITIONS, INFRASTRUCTURE MANAGEMENT, AND LAND-USE MANAGEMENT

Facing big issues of population decline and a hyper–aging society in Japan, the government united to deal with the issues, and established in the Cabinet the Headquarters for Overcoming Population Decline and Vitalizing Local Economies (the HQs) in Sep. 2014 so that each region could create an autonomous and sustainable society utilizing its own characteristics. (The establishment of this headquarters was decided in the Cabinet meeting on Sep. 3, 2014.) Prior to this establishment, the Ministry of Land, Infrastructure, Transport and Tourism (MILT) announced the Grand Design of National Spatial Development towards 2050, Japan—creation of a country generating diverse synergies among regions, and disclosed its vision for the Grand Design to respond to key issues such as: sharp population decline; declining birth rate; advancement of an unprecedented hyper-aging society; approaching a huge disaster; obsolete infrastructure; restriction of food, water, and energy; and global environmental problems and others. Also, among the comprehensive strategies (decided in the Cabinet meeting on Dec. 26, 2014), the HQs proposed the formation of small core areas in the rural regions and reinforced management of the existing stock with regard to land-use management. For these issues that the government proposed, we would like to discuss in this chapter interpretation of the proposed policies, and the direction toward a solution from the viewpoint of a researcher.

4.6.1.1 Changes in the Socio-Economic and Fiscal Conditions

Among people who study land-use management today, it is a commonly held idea that we are under very severe socio-economic and fiscal conditions, and that resilience will also be decreased unless some actions are taken. First of all, we shall study this point from the macro viewpoint.

Disaster Resilient Cities
http://dx.doi.org/10.1016/B978-0-12-809862-2.00017-6

Let us see the number of victims due to disasters. The number is an index of resilience. From the trends after 1960, we see that the number of victims in the 1960s is larger than that in any other decade, but there is no apparent tendency for improvement from the 1970s up to the present, even when we exclude the victims of the two great disasters, the Great Hanshin-Awaji Earthquake and the Great East Japan Earthquake (Fig. 4.6.1). During these periods, despite continuous investments in disaster measures, a huge amount of study on disaster prevention and mitigation, and various enactments for construction standards and city planning, there has not been any apparent improvement of resilience as far as we can see in the number of the victims, and we must realize that Japan is still a country vulnerable to disasters.

The Grand Design of National Spatial Development towards 2050 warns that division of the Japanese land into $1\,km^2$ mesh shows that in the year 2050, there will be no people living in 19% of the meshes, and the population will be reduced to less than half in 44% of the meshes (Fig. 4.6.2). As to the countermeasures for population decline in regional areas, the HQs is now discussing various measures from three basic viewpoints: (1) realizing the young generation's wishes for employment, marriage, and child-care; (2) stopping the "over concentration in Tokyo";

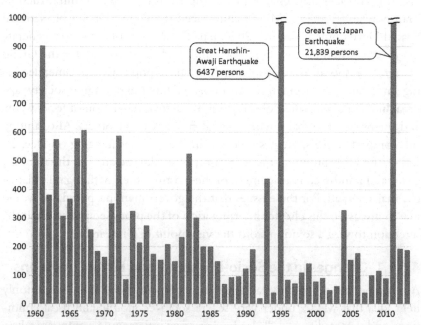

Fig. 4.6.1 Trends of the number of dead and missing due to natural disasters (unit: person). *(Data from Cabinet Office, Government of Japan)*.

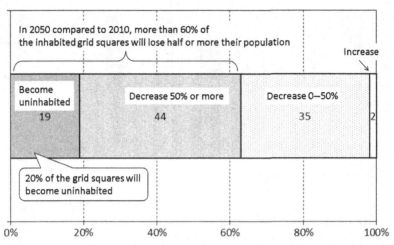

Fig. 4.6.2 Ratio of the number of meshes by population increase/decrease in 2050 (basis year: 2010). *(Data from MLIT).*

and (3) solving regional issues utilizing their characteristics. However, it will take some time for the effects to be materialized, and the critical conditions will continue for some time.

Let us have a look at the fiscal condition. According to the Ministry of Finance (hereafter referred to as MOF), the total amount of culture, education, and science promotion expenses, defense-related expense, and public works-related expenses accounted for 52.5% of the total expenditure in the fiscal year 1980, but dropped to 27% in fiscal year 2014 (Fig. 4.6.3). This means that the share of the government investment for future generation was decreased by half. Furthermore, according to the MLIT, most of the infrastructure that was intensively built up during the rapid economic growth period is now 30–50 years old; therefore, it is expected that it will rapidly become obsolete and need repair and reconstruction. Taking the roads and bridges as an example, the percentage of the infrastructure that is more than 50 years old was 8% in fiscal year 2012, but will jump to 53% in fiscal year 2030 (Fig. 4.6.4). Based on capital spending in the past, the MLIT estimated the future maintenance, management, and renewal cost that will be required for the infrastructure under their jurisdiction (roads, port facilities, airports, public rental housing, sewage, city parks, flood control, and coasts). According to their estimate (Fig. 4.6.5), the future maintenance cost and renewal cost of old facilities only will be almost equal to the current public investment amount. It is expected that even the maintenance and renewal of existing facilities will become difficult, not to mention the preparation of new facilities.

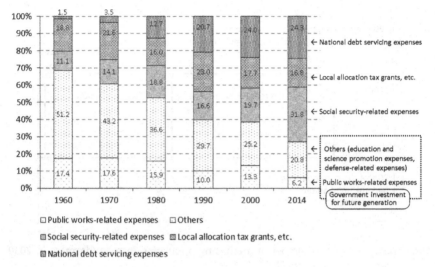

Fig. 4.6.3 Trends in Japan's fiscal condition. *(Data from MOF).*

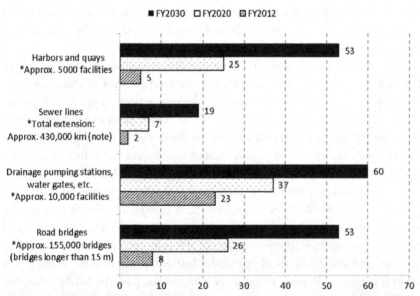

Fig. 4.6.4 Percentage of the infrastructure more than 50 years old (unit: %). *(Data from MLIT).*

As noted earlier, if the current socio–economic and fiscal conditions continue, land-use and infrastructure management will become very difficult sooner or later, and the resilience of Japan will be inevitably lowered more and more with the increasing external forces that will be described later.

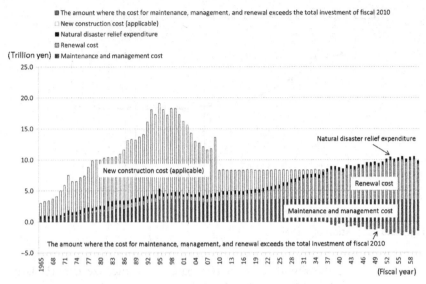

■ The amount where the cost for maintenance, management, and renewal exceeds the total investment of fiscal 2010
□ New construction cost (applicable)
■ Natural disaster relief expenditure
▨ Renewal cost
■ Maintenance and management cost

Fig. 4.6.5 Estimated amount of infrastructure, assuming the same maintenance, management, and renewal as in the past. *(Data from MLIT).*

4.6.1.2 Increasing Vulnerability Caused by Urban Growth and Sprawl

The Japanese land is not a great plain of a continental type, but a very complicated terrain of mixed mountains, plains, rivers, and coastlines. Therefore, from olden times people have avoided living in places that are vulnerable to natural disasters. However, since the latter half of the 1900s, due to the rapid expansion of urban areas which was brought about by the increase of urban population, people have started to live in the places where there were neither people living nor industries in olden times. The comparison of the year 1960 and the year 2010 shows that the densely inhabited district (DID) area in 2010 was approximately 3.3 times larger (Fig. 4.6.6). We cannot deny that this rapid expansion has brought about greater vulnerability to disasters, and it is one of the reasons for increased natural disasters in recent years, together with more violent weather.

The administration has taken some countermeasures during this period against the increase in causative factors of disasters due to the expansion of urban areas. With the rapid growth of the Japanese economy, the City Planning Act was enforced in 1968, by which development control of areas vulnerable to disasters was introduced by the demarcation of urbanization areas and urbanization control areas, and other development regulations. However, there are two kinds of limitation in the demarcation. First, in

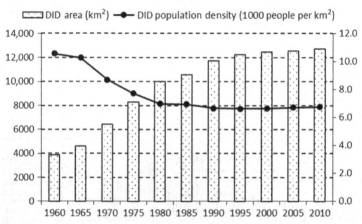

Fig. 4.6.6 Expansion of DID area.

the areas where the disaster danger is apparent and the size is fairly large, the danger can be well considered when the demarcation is conducted. Meanwhile, in the locally dangerous places or in areas where danger signs are not tangible, we do not think that sufficient consideration has given to demarcation. Fig. 4.6.7 shows overlapping of landslide-danger zones and urbanization areas. The other kind of limitation is how to respond to the changes of external forces after the demarcation. As in the flood near Hakata station in 1999 (Photos 4.6.1 and 4.6.2), and the landslide in Hiroshima City in 2014, there have been many cases in which newly developed areas are hit by external forces, such as torrential rain, that were not so tangible when the areas were developed. When we compare the observed number of times that heavy rains exceeding 80 mm/hour hit 1000 points, the year 2010 is nearly two times that of 1975 (Fig. 4.6.8).

In these circumstances, it has been a long time since the necessity of expanding nonstructure based (soft) countermeasures was proposed, such as alarms for evacuation to protect human life. For various alarms frequently raised during heavy rains and floods, the administration had not necessarily issued proper evacuation recommendations and instructions to residents. Recently, due to the instructions of the government, local governments of cities, towns, and villages have come to make use of the government's evacuation recommendations and instructions; however, still not all residents evacuate as recommended or instructed. According to the Sankei News (Oct. 20, 2014), on Oct. 6 when typhoon no. 18 hit downtown Tokyo, Minato ward of Tokyo issued an evacuation recommendation to about 23,000 households, or 45,000 people, but only six people went to two

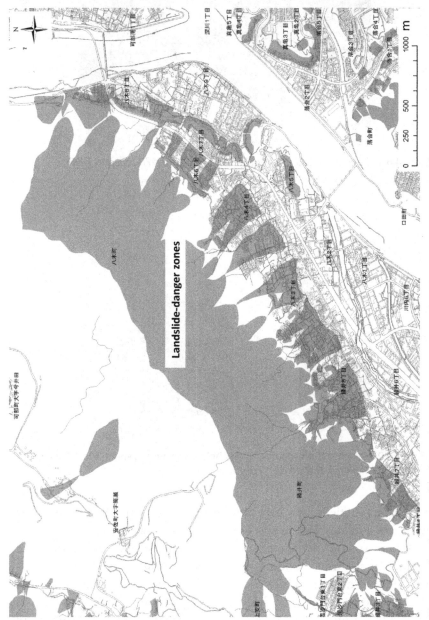

Fig. 4.6.7 Overlapping of landslide-danger zones and urban areas (Hiroshima City).

Photo 4.6.1 Flood near Hakata station in 1999. *(Data from Flood report 1999).*

Photo 4.6.2 Hakata station and the vicinity before the development. *(Data from Report of land readjustment in Hakata area, Fukuoka).*

evacuation facilities to stay. This illustrates that an evacuation recommendation is not a mighty tool to save lives from natural disasters.

When we consider the external forces that would exceed the level assumed at the time of development, and the difficulty that nonstructure countermeasures such as evacuation recommendation would not look perfect, it is very difficult to secure the safety of residents as long as we continue the current land use that many people reside in disaster-prone areas. However, there have not been any cases to review the demarcation of urbanization areas and urbanization control areas

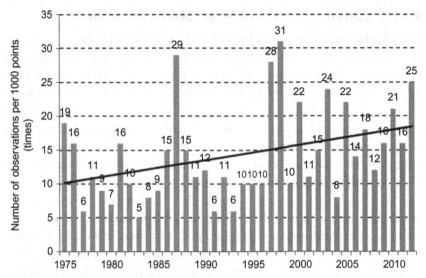

Fig. 4.6.8 Observed number of times per year of heavy rain exceeding 80 mm/hour (Automated Meteorological Data Aquisition System (AMeDAS)). *(Data from Japan Meteorological Agency (JMA)).*

where disaster risk has become higher compared with that of the time of original demarcation due to the change of external forces and level of safety of the area.

4.6.2 POLICY REFORMS/MEASURES IN PROMOTING SMART SHRINK STRATEGY IN JAPAN

"Smart Shrink" is the opposite of "Smart Growth" which means growth control, aiming to form a sustainable area. It is a general concept to refer to the regional management technique for maintaining and improving the quality of life (QOL) of residents under absolute population decline. It means that "a region must shrink wisely" by finding its own characteristics and securing competitiveness against other regions, while trying to make public enterprises efficient or providing efficient public services [1]. Smart Shrink has many similarities with compact city policy in that Smart Shrink has to make the residential area compact. In order to secure land resilience in the future, it is necessary to strategically make residential areas more compact not only in urban areas but also in rural regions. In this section, we shall discuss and clarify policy reforms or measures under the current system to promote Smart Shrink regarding securing resilience against disasters.

4.6.2.1 Current Situation of Smart Shrink Promotion

Considering our future population decline and the tight fiscal conditions, the finances of local governments will become more and more difficult. Furthermore, if we continue the current land use and living patterns, the maintenance cost of the infrastructure will not be reduced much, laying a big burden on local governments, which will lead to the deterioration of local governmental financial conditions. There have been some activities to promote making residential areas compact by the compact city policy, etc., as written in the second report of the panel on infrastructure development [2].

If we check regional renovation cases of the MLIT database, we shall find many cases of the compact city policy that have taken effect mainly in the heavy-snow areas of the Tohoku and Hokuriku districts (Fig. 4.6.9). For example, according to the snow-removal work plan of Aomori City, the annual snow-removal cost of the city is more than twice that of the mainte-nance and control cost for public works, and it is a very big financial burden; therefore, it is an urgent issue to make snow-removal more efficient by the compact design of urban areas (Fig. 4.6.10). Taking this into consideration, in areas where the snow-removal cost is huge, it is quite possible that there will be a fiscal-economic rationality in promoting the Smart Shrink strategy.

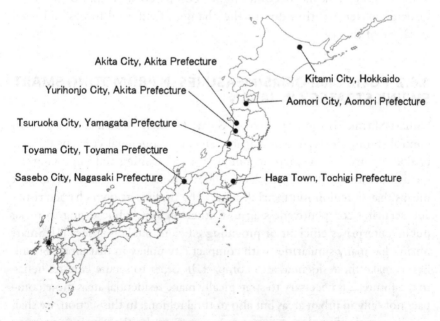

Fig. 4.6.9 Examples of local governments implementing the compact city policy. (*Data from MLIT*).

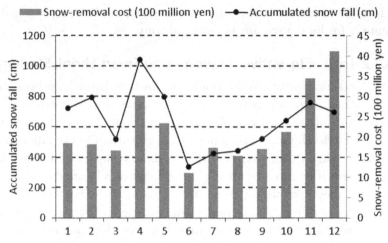

Fig. 4.6.10 Snow-removal cost and accumulated snowfall. *(Data from Aomori City).*

The Low Carbon City Act is a law that aims clearly at making residential areas compact. However, only 16 local governments had made low carbon city plans according to the Low Carbon City Act as of Nov. 1, 2014, and there is not yet a widespread movement (Fig. 4.6.11). It is difficult to promote low carbon city plans and proper location plans unless there is some concrete merit such as improvement of the living environment and economic merit for the residents, as well as fiscal merit for

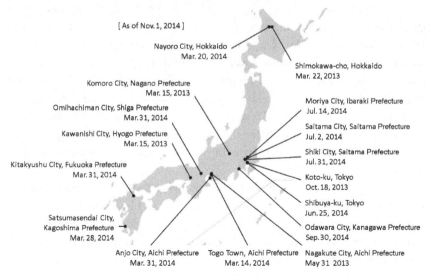

Fig. 4.6.11 Progress of the low carbon city plans by the Low Carbon City Act. *(Data from MLIT).*

the local governments. In the following sections, we shall discuss possible incentives for local governments and residents.

4.6.2.2 Lack of Incentives in Promoting Smart Shrink Strategy

So far, land-use regulations for residential areas in Japan have been primarily aimed at newly developed/located areas, and the demarcation between urbanization areas and urbanization control areas, and the restrictions by development restrictions were effective. On the other hand, in order to make existing residential areas more compact, relocating current residents is required, but it is quite difficult to relocate them by regulations as they have already permitted to reside there. For example, the government can advise people in the landslide special alarm areas to move to other places by restrictive measures, but there had not been such a case of a moving advice issued to the residents in those alarm areas as of Aug. 31, 2005, according to the MLIT. When such residents moved they were not motivated by such restrictive measures as moving advice, but by incentive-guided measures such as subsidies and grants. As noted earlier, restrictive measures are not effective for existing residents, and incentives such as subsidies and grants are more effective. However, currently, subsidies and grants for existing residents are insufficient in ordinary areas except the Special Financial Support for Promoting Collective Relocation for Disaster Mitigation.

The Kumamoto Daily News said in its morning edition of Nov. 9, 2013 that the results of the survey conducted by Aso City regarding housing in the future for a total of 523 households in 14 administrative districts in the Ichinomiya district of Aso City (344 households, approximately 66% of the total responded), where houses were severely damaged due to the landslide caused by torrential rain that hit northern Kyushu in 2012, showed that out of the 101 households whose houses were totally or half damaged, 46 households (that is, nearly half) responded that they wished to move somewhere, saying that the subsidy for the land and housing was the biggest issue. Here, we can see the importance of some assistance measures regarding relocation.

In order to realize Smart Shrink for securing resilience against disasters, it is essential to move from disaster-danger areas. However, as mentioned previously, the important point for realizing Smart Shrink is whether subsidies or grants are financially feasible or not, and it is not well prepared at this moment.

4.6.2.3 An Incentive for Local Governments to Promote the Smart Shrink Strategy

Main administrative bodies to advance Smart Shrink are local governments such as cities, towns, and villages. Currently in Japan, most local governments cannot cover their expenditures by their tax revenue. More than 50% of their income comes from grants and subsidies from the national government. The main financial assistance to local governments from the national government is the ordinary local allocation tax. We shall discuss whether the current fiscal assistance system helps to promote Smart Shrink.

The ordinary local allocation tax is calculated by the following equation.

$$\begin{aligned}\text{The amout of ordinary local} &= \text{The amount of standard}\\ \text{allocation tax} &\quad \text{financial need}\\ &\quad -\text{The amount of standard}\\ &\quad \text{financial revenue}\\ &= \text{Shortage of financial source}\end{aligned}$$

As shown in Fig. 4.6.12, the larger the standard financial need, the larger the amount of ordinary local allocation tax. In other words, if standard financial need is reduced, the amount of ordinary local allocation tax is reduced; therefore, there is no financial incentive for local governments to reduce the standard financial need.

The amount of standard financial need is calculated by an individually set measuring unit for each of the administrative items such as the extended

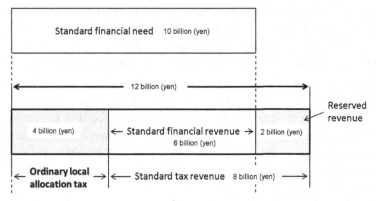

Fig. 4.6.12 System of ordinary local allocation tax. *(Data from Ministry of Internal Affairs and Communications (MIC)).*

Table 4.6.1 Examples of the items for calculating the amount of standard financial need and their measuring units

Civil engineering works		Education	
Item	Measurement unit	Item	Measurement unit
Road and bridge	Area of road	Elementary	Number of children
	Length of road	school	Number of classes
Port and harbor	Length of mooring facility		Number of schools
	Length of protective facility	Junior high school	Number of students
Urban planning	Population in city planning area		Number of classes
Park and green space	Population		Number of schools
	Area of urban park	High school	Number of teaching staff
Sewage	Population		Number of students
Other	Population	Other	Population
			Number of children of kindergarten

distance of roads or the number of schools. The example of those items and measuring units are shown in Table 4.6.1.

Table 4.6.1 shows that the larger the assets, such as the extended distance of roads, or the number of schools, the larger the amount of standard financial need is calculated to be. If local governments realize a Smart Shrink policy, the amount of standard financial need will decrease as the existing assets will be reduced. On the other hand, under the current system, reducing the amount of standard financial need leads to reducing the amount of ordinary local allocation tax. So, there will be no financial incentive for cities, towns, and villages to reduce existing assets.

Meanwhile, the MLIT has published the Evaluation handbook of urban structure [3] and in this handbook, "the population density in the area along public transportation lines," "the population ratio in the disaster-danger area," "the administrative cost per citizen," etc., are listed as the evaluation indices of urban structures. The MLIT has given high marks to the cities that are emitting less CO_2 due to traffic, resilient to disasters, and highly efficient in the city administration, and has introduced policies to give financial assistance to those cities.

As its concrete example, the Act on Special Measures concerning Urban Renovation was revised in 2014 (Fig. 4.6.13). The Act says that local governments can make their own location optimization plan. By this plan,

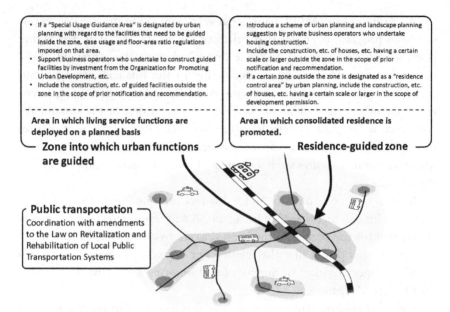

Fig. 4.6.13 Revision of the Act on the Measures concerning Urban Renovation and Vitalization; city functions/setting of the areas for relocation promotion. *(Data from MLIT).*

government financial assistance can be given to the cities, towns, and villages, and this assistance will become a big incentive to them.

Smart Shrink policies cannot be popular among the existing residents, since relocation will bring changes to their lives, and extra costs. Also for local governments, there is not much financial incentive, as explained in the previous. So, it is difficult to promote Smart Shrink policies under the current fiscal system. In this sense, fiscal policy changes are needed.

4.6.2.4 Discussion on Measures of Location Management Toward Smart Urban Shrinkage in Disaster Dangerous Zones

As a form of assistance for local governments during disasters, there is a government subsidy system by the disaster reconstruction program. The disaster reconstruction program reconstructs public facilities such as roads, rivers, and schools, or farmland and agricultural facilities when they are damaged due to natural disasters by typhoon, heavy rain, earthquake, etc. There are two kinds of program; one is carried out under direct control of the government, and the other is conducted by the local governments receiving subsidy from the government. (Some programs are conducted by private sectors such as railway companies.)

According to the MLIT, national government aid in the disaster reconstruction program occupies two-thirds of the entire cost, which is higher that the ordinary program. In some test calculations, the actual aid of the national government occupies 98.3% due to the financial arrangements using local allocation tax. Comparing with the ordinary national government aid of general public works (50%), fiscal burden of the disaster reconstruction program is very low (Fig. 4.6.14). The basic rule of disaster reconstruction programs is to "reconstruct to the original state" by reconstructing public facilities such as damaged roads to their condition before the disaster.

In areas that are frequently hit by disasters, it is quite often that the facilities once hit and recovered are hit again. So, to prevent repeated disasters, it is desirable for the people in the areas that are frequently hit by disaster to relocate to safer places after the disaster. For this system, there is a program called the Special Financial Support for Promoting Collective Relocation for Disaster Mitigation.

For local governments, however, the financial burden of this program is bigger than that of the disaster reconstruction program. Even if it is desirable that the people living in damaged houses are relocated to houses in safe areas through the relocation program, there are many cases where

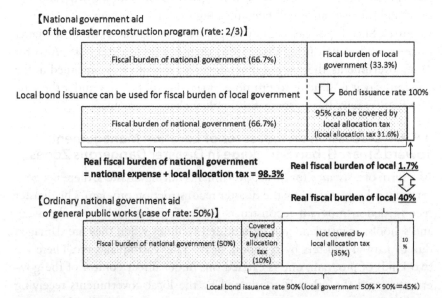

Fig. 4.6.14 High rate of the national government aid in the disaster reconstruction program. *(Data from MLIT).*

local governments choose "reconstruction to original state" by the disaster reconstruction program, which is financially less expensive. Also, for the people living in damaged areas, since they must bear most of the relocation costs of their houses and private assets, the burden is too big to shoulder without any public assistance. However, considering the situation in which disasters are often caused by larger external forces due to climate change, "reconstruction to original state" may not be sufficient to prevent future disasters.

In the next section, we shall study whether the Special Financial Support for Promoting Collective Relocation for Disaster Mitigation can be carried out in case the restrictions of the above-mentioned financial assistance system are loosened in order to advance the Special Financial Support for Promoting Collective Relocation for Disaster Mitigation.

4.6.2.5 Feasibility of Implementation of Group Relocation Project for Disaster Prevention by Fiscal Arrangement Modification

The areas susceptible to landslides and other disasters are on the edges of residential areas, and these are the areas that can be the subjects of Smart Shrink. By relocating to the nearby main community as planned, people can secure safety against disasters, and we can expect a reduction of the cost of disaster reconstruction in the future since these areas will become nonresidential and no need to reconstruct local residential infrastructure.

We would like to introduce a test calculation [4] of the Kyushu region to show as an example of whether it is financially feasible to relocate to the nearby community from the landslide-danger zone. We calculated the total amount of (1) the landslide disaster reconstruction expenses (public works facilities, general assets, and public engineering works) that will be reduced by relocation, and (2) the infrastructure maintenance expenses (local roads, water supply, and sewer infrastructure) that will not be required after the relocation. Then we checked whether the expense reduction ((1) and (2)) will cover the expenses (land purchase cost of the relocated area and the subsidy for the housing preparation cost in the relocated area) that are needed for relocation. We supposed that the test calculation period was 50 years, and the discount ratio at 4% by which the expenses arising in the future can be converted to the current value. Furthermore, relocation was limited to the nearby mother community. (Long-distance relocation was not considered (Fig. 4.6.15).) The entire Kyushu region was divided into areas, with each

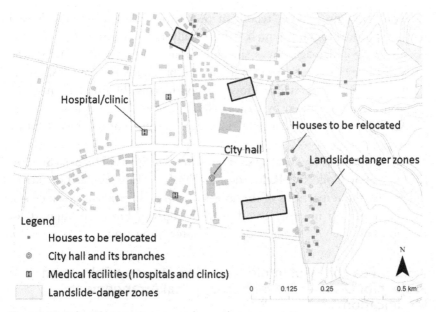

Fig. 4.6.15 Relocation image to a nearby mother community.

area size $500\,m \times 500\,m$, and the relocation destination was limited to the neighboring area (relocation distance is 500–1500 m). The results of the test calculation under these conditions are shown in Fig. 4.6.16.

The number of areas where the financial balance improves by relocation is 8264, approximately 18% of the total number. Those improved areas are mostly located in rural regions.

The number of the areas where the financial balance improves for each mesh (500 m mesh) is standalone. However, if we aggregate each relocation meshes balance into local government constituency and check financial balance by a local government basis, the number of areas that can be relocated in all of Kyushu will be 15,573 areas, or 35% of the total. The areas that can be relocated are shown in Fig. 4.6.17.

The above test calculations shows that by improving the current disaster reconstruction program (the target is only public facilities, and that is based on the principle of reconstruction to original state), Smart Shrink would be promoted and security of residents could be achieved.

4.6.2.6 Future of Japanese National Land Without Implementation of Smart Shrink

So far, disaster reconstruction programs have been supported by great financial assistance from the government. However, Japan's financial situation will become more and more severe in the future. Sato et al. [5] point out that

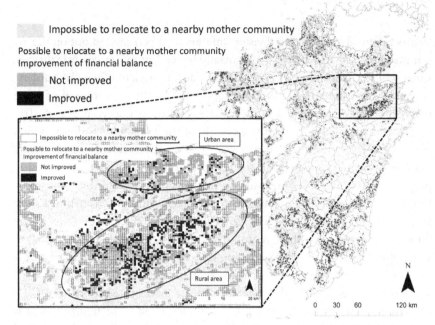

Fig. 4.6.16 Test calculation results of the areas of which financial balance improves by the relocation to a nearby mother community (for each district).

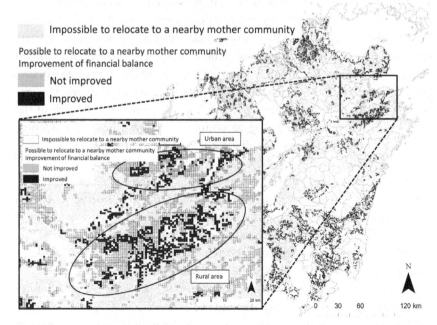

Fig. 4.6.17 Test calculation results of the areas for which the financial balance improves by relocation to a nearby mother community (when the total financially improvement for each district covers the total of financial loss for each district).

Japan's financial assistance to disaster sufferers has been carried out on the assumption that Japan will not be financially broke, but in a large-scale disaster, Japan will not be in a position to shoulder an unrestricted financial burden.

Supposing the current generous support from the government of more than 98% were changed to burden sharing of local governments of 10%, what will happen? Let us see the example of the Takachiho Railway, which was operated in the northern part of Miyazaki Prefecture. It was a private company, so the company had to bear 50% of the disaster reconstruction program, and could not afford to pay the expenses needed for reconstruction from a flood in 2005, and was obliged to withdraw from the railway business. If aging and the shrinkage of regional economies proceed in the future, the capability of these regions to share expenses will further decrease. Once these regions are struck by a big disaster, they may not be able to bear the cost of disaster reconstruction for their regions, and regional financial shortfall may lead to a "sudden death" of local government operation.

When we consider the above situation, we need to recognize that it is impossible both from the point of disaster protection and from the point of finance to secure resilience to disasters if we to continue the current land use and social infrastructure as they are.

If the percentage of regional burden in the disaster reconstruction program were to increase, it would bring a great amount of financial burden to local public organizations. In southern and western Japan especially, there are a lot of heavy rain disasters, and almost every year, those regions are hit by terrible disasters. As a result, enormous amount of disaster reconstruction expense have to be used each time. The damage amount of public facilities is not proportional to the size area, but depends greatly on the size of the residential area. If we are going to continue to keep the vast residential area in disaster-prone area, the disaster reconstruction expenses will even increase by the growing disaster external force.

We checked yearly disaster reconstruction program expenses in northern Kyushu. The expense for the "unprecedented" 2012 northern Kyushu torrential rain was an enormous amount of 35 billion yen. However, it was found that the disaster, of which expense amounts to more than 60% of that "unprecedented disaster expense," occurs once every 2–3 years. In northern Kyushu, the average yearly expense over the past 10 years is almost equal to the maintenance and management expense for the public infrastructures such as roads and water supply (Fig. 4.6.18).

The principle of disaster reconstruction program is "reconstruction to original state." Therefore, even if the disaster damaged is restored, it is quite

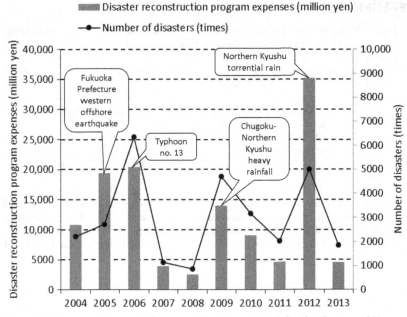

Fig. 4.6.18 Disaster reconstruction program expenses in Fukuoka, Saga, and Nagasaki prefectures. *(Data from MOF).*

likely that the place will be damaged again if the bigger disaster external force hits there. Considering that some of the heaviest rains on record have occurred many times now, it is expected that huge disaster reconstruction program expenses will be continuously required.

Suppose a small village of a few households located on the edge of a community in the heavily damaged area were moved after the damage to the nearby mother community, which was comparatively safe, rather than being reconstructed to the original state; the reconstruction expenses needed to protect the edge part locally will not be required. This would enable us to reduce the facilities that might otherwise have required continuous reconstruction expenses to return to the original state after possible future disasters, and would lighten the financial burden in the future. In this way, by focusing on disaster reconstruction program expenses, we can see that improvement in the regional maintenance cost by the Smart Shrink strategy can also be applied to areas with frequent disasters as well as heavy-snow areas. Furthermore, by moving peripheral areas that are frequently damaged by landslides to the safe nearby mother community, we can improve not only disaster protection capability but also access to various facilities such as commercial, medical, and welfare facilities that are needed in daily life, and enhance the QOL.

REFERENCES

[1] Hayashi Y. The only strategy for urban survival—"Miserable Shrink" or "Smart Shrink". Monthly news of "The Gifu Chamber of Commerce and Industry"; 2014. p. 2–6 [in Japanese]

[2] City and Regional Development Bureau, Ministry of Land, Infrastructure, Transport and Tourism (MLIT). Toward realization of Urban Renovation and compact cities. 2007 [in Japanese].

[3] Ministry of Land, Infrastructure, Transport and Tourism (MLIT). Development of evaluation handbook of urban structure. 2014 [in Japanese].

[4] Kajimoto R, Kachi N, Tsukahara K, Akiyama Y. Financial feasibility study on residential optimization at village level around disaster danger zone. In: Proceedings of infrastructure planning, vol. 49. Japan Society of Civil Engineers, Tokyo; 2014. CD-ROM [in Japanese].

[5] Sato M, Miyazaki T. Intergovernmental risk allocations and the budget for the reconstruction from the Great East Japan Earthquake disaster. In: Local government finances—risk sharing between governments. Financial Review, vol. 108. Ministry of Finance Japan, Policy Research Institute, Tokyo; 2012. p. 33–53 [in Japanese].

[6] Cabinet Office, Government of Japan. White paper on disaster management 2014. 2014 [in Japanese].

INDEX

Note: Page numbers followed by *f* indicate figures and *t* indicate tables.

Printed in the United States
By Bookmasters